世纪互联蓝云研究院丛书

PowerShell for Office 365 应用实战

世纪互联蓝云公司 主编

蔡南 杨好骥 李阳 陈飞 编著

电子工业出版社
Publishing House of Electronics Industry
北京·BEIJING

内 容 简 介

本书选取企业日常信息管理中出现频率较高的应用场景，采用循序渐进的讲解方式，指导读者使用 PowerShell 来配置和管理 Office 365 的应用与服务。同时，通过对比介绍图形管理界面相应的操作步骤，来降低学习的难度。书中结合真实的 PowerShell 应用实例进行讲解，并给出了部分命令常见的报错输出信息，有助于提高读者的命令使用与排错能力。

全书共包括 5 章。第 1 章介绍了使用 PowerShell 管理 Office 365 的基础知识和优势，以及使用 PowerShell 管理 Office 365 及其应用需要进行的准备工作；第 2 章讲解了使用 PowerShell 管理 Office 365 的租户的方法；第 3 章到第 5 章分别讲解了使用 PowerShell 管理 Office 365 的 Exchange Online、Skype for Business Online 和 SharePoint Online 的相关知识。

本书特别适合 Office 365 的管理员学习使用。另外，广大程序开发人员、追求进一步提升技术能力的网络管理人员及大专院校计算机专业学生阅读本书，也能从中获益。

未经许可，不得以任何方式复制或抄袭本书之部分或全部内容。
版权所有，侵权必究。

图书在版编目（CIP）数据

PowerShell for Office 365 应用实战 / 世纪互联蓝云公司主编：蔡南等编著. —北京：电子工业出版社，2020.10
（世纪互联蓝云研究院丛书）
ISBN 978-7-121-37197-4

Ⅰ. ①P… Ⅱ. ①世… ②蔡… Ⅲ. ①办公自动化—应用软件 Ⅳ. ①TP317.1

中国版本图书馆 CIP 数据核字（2019）第 168450 号

责任编辑：张瑞喜
印　　刷：中国电影出版社印刷厂
装　　订：中国电影出版社印刷厂
出版发行：电子工业出版社
　　　　　北京市海淀区万寿路 173 信箱　邮编：100036
开　　本：787×1092　1/16　印张：15.75　字数：403 千字
版　　次：2020 年 10 月第 1 版
印　　次：2020 年 10 月第 1 次印刷
定　　价：55.00 元

凡所购买电子工业出版社图书有缺损问题，请向购买书店调换。若书店售缺，请与本社发行部联系，联系及邮购电话：（010）88254888，88258888。
质量投诉请发邮件至 zlts@phei.com.cn，盗版侵权举报请发邮件至 dbqq@phei.com.cn。
本书咨询联系方式：zhangruixi@phei.com.cn。

序　言

脚本语言是计算机编程语言的一种，源于早期计算机系统的文字模式的命令行。为了减少不断重复的命令行的文字性输入，通过批处理的方式把这些文字性命令行汇聚到一起形成按批次执行的命令脚本。从服务器操作系统历史来看，UNIX 系统一直有着功能强大的脚本（Shell）。脚本对于系统管理员来说，有着无可比拟的优势，便于他们高效地进行操作系统的管理。而 Windows Server 在诞生之初，是以图形化操作界面为主的，这种图形化的管理方式便于初学者上手，但是却不利于资深管理员进行批量及深入的管理。虽然微软有 WMI 这个应用程序

世纪互联蓝云 CTO　汤　涛

接口用来管理 Windows 操作系统，而且 WMIC（WMI 的命令行接口）功能强大，但是它的语法让很多人觉得难以使用。为此，微软启用了代号为 Monad 的项目。Monad 字面的意思是"单一"，顾名思义，Monad 的意图就是做出一套整合完善的脚本语言工具，让使用者不需要再用许多不同的工具。在 Monad（PowerShell 的前身）发布后的第二年（2006 年），微软将其改名为 PowerShell 1.0。PowerShell 捆绑.NET，借助.NET Framework 平台强大的类库，不仅能访问.NET CLR，而且可以使用现有的 COM 技术，在进一步扩展 Windows 命令和 Windows Script Host 的基础上，实现强大的管理功能。从此 PowerShell 发展成为一个脚本平台，很多成型的微软产品都有对应的 PowerShell 的应用程序接口，如 SQL Server 和 Microsoft Azure。

Office 365（现更名为 Microsoft 365，本书沿用 Office 365 这一名称）是微软的新一代云计算产品，包括了微软 Office 套件，即 Word、Excel、Outlook、OneNote 和 PowerPoint，并集成了快捷的电子邮件 Exchange Online、实时的文件共享 SharePoint Online 和稳定的视频联机会议 Skype for Business Online/Teams 等先进的企业级云服务。自 2014 年 4 月由世纪互联运营的 Office 365 落地中国以来，Office 365 已经服务于中国超过 3 万家企业的 240

多万用户；Office 365 企业版月活跃用户超过 1.55 亿户；90% Office 新增中国客户选择 Office 365 云服务，Office 365 已经成为全球众多企业及组织机构运营和扩展业务的重要工具。

世纪互联蓝云作为 Microsoft Azure、Office 365、Dynamics 365 和 Power Platform 云服务在中国的运营平台，多年来，一直秉承高技术、高可靠、高质量的服务，并提供企业级服务等级协议保证。在长期的运营服务过程中，世纪互联蓝云的工程师们在深入了解了用户在 Office 365 日常应用管理方面可能会遇到的一些棘手的情形后，发现很多状况在 PowerShell 强大的功能面前都会有很好的解决办法。为此，世纪互联蓝云的工程师们利用工作之余，总结并撰写了此书。此书凝结了世纪互联蓝云的工程师们长期以来在 Office 365 上的宝贵经验，如今将之分享出来，以期帮助更多的 Office 365 使用者更好地应用这一工具。通过本书，也期待更多的 PowerShell 爱好者们加入 Office 365 的大家庭中，使 Office 365 的应用更百花齐放。

世纪互联蓝云 CTO　汤　涛

2020 年 8 月

前　言

经过一年的努力和筹备，这本《PowerShell for Office 365 应用实战》终于和大家见面了。为了这本书，我们几个参与撰写的工程师可真是费了不少力气呢！从最初的选题和编排，到后来的撰写与修改，我们都绞尽脑汁，有时甚至为了某个场景的展现争得不可开交。归根结底，都是为了使大家可以从本书中得到一点启发，或是找到一些可以借鉴的内容，这样就足够了。

本书围绕企业日常信息管理中出现频率较高的应用场景展开，介绍使用 PowerShell 来配置和管理 Office 365 的应用与服务的相关知识。

第 1 章主要介绍了使用 PowerShell 管理 Office 365 的基础知识。包括什么是 PowerShell，使用 PowerShell 管理 Office 365 及其各个应用所需进行的准备工作，注册 Office 365 测试账户的具体步骤，相关组件的下载安装等。

第 2 章主要讲解如何使用 PowerShell 管理 Office 365 的租户。包括管理租户中的用户、安全组、角色和订阅的具体方法，以及需要注意的问题。

第 3 章主要讲解如何使用 PowerShell 管理 Office 365 的 Exchange Online。具体内容包括连接 Exchange Online 的步骤，连接过程中可能遇到的问题，邮箱、组等 5 种不同类型的收件人的管理，以及管理员角色、用户角色、OWA 策略等 Exchange Online 权限的管理。

第 4 章主要讲解如何使用 PowerShell 管理 Office 365 的 Skype for Business Online。具体内容包括连接 Skype for Business Online 的步骤和可能遇到的连接问题，用户及其属性的查看和用户属性的修改，以及租户的常规属性与外部通信属性的查看和修改。

第 5 章主要讲解如何使用 PowerShell 管理 Office 365 的 SharePoint Online。具体内容包括网站及其属性的查看和修改、网站的添加和删除、查看和管理网站中的组、查看和管理网站中的用户，以及 OneDrive for Business 个人网站的管理。

参与编写工作的工程师都来自世纪互联蓝云，世纪互联蓝云是世纪互联成立的全资子公司。作为微软云服务在中国的提供方，世纪互联蓝云为广大用户提供了具有财务保障的高达 99.9% 的月度满意度服务等级协议保证，以及实时公开的服务仪表盘等措施，让公众实时监督其可信性和安全性，并且还累计获得了云主机、对象存储、云数据库、云引擎、全网负载均衡、云备份、企业电子邮件、文件共享协作、联机会议、混合云解决方案（公有云）、混合云解决方案（私有云）等 11 项可信云认证。由世纪互联运营的 Office 365 获得了可信云 2014—2015 年度行业云服务奖的办公应用奖；由世纪互联运营的 Microsoft Azure 获得 2015 年度最佳业务连续性策略实施奖。世纪互联蓝云，2016 年凭借运维安全可靠的 Microsoft Azure 和 Office 365 云服务荣获 2016 年度中国数据中心云运维首选品牌奖；2017 年凭借 Azure 混合云解决方案获得可信云技术创新奖之混合云奖及混合云可信云认证；2018 年获得中国云计算领域值得信赖品牌和中国云计算领域最佳解决方案奖；2019 年荣获年度优秀云计算服务商奖。

参与本书编写的全体成员是：蔡南、陈飞、李阳、杨好骥和叶海燕。其中，第 1 章和第 2 章由蔡南负责编写，第 3 章由蔡南和陈飞共同编写，第 4 章由蔡南和李阳共同编写，第 5 章由杨好骥和蔡南共同编写，全书由蔡南进行统稿和完善，叶海燕负责本书序言、前言的组织编写和全书整体形象审定。

在此，感谢世纪互联蓝云 Office 365 团队的同事在书籍出版过程中给予的支持与帮助。另外，还要特别感谢以下同事的鼎力支持：程文中、刘钧、张旭、杨勇刚、张弛、马洪英、王少锋、艾青、王超、王宏友、陈亚群、吴超、马俊和杨振强。

由于 PowerShell 涉及的知识点比较多，虽然我们尽力反复推敲、力求准确，但是书中难免会有不妥或错误之处，非常欢迎大家与我们进行联系和交流，共同完善疏漏之处。有关本书的任何问题、意见或建议，可以发送邮件到 O365Book@oe.21vianet.com，与我们进行联系。

最后，感谢您关注由世纪互联运营的 Office 365，并再次感谢您选择本书！

蔡 南

2020 年 8 月

如何使用本书

如果您已经有一些使用 PowerShell 的基础，那本书就是为您量身定做的。您现在就可以选取自己感兴趣的内容，并开始 PowerShell for Office 365 探索之旅了。如果您没有任何使用 PowerShell 的经验，您可以先参考附录 A 部分，学习一些与本书中出现的 PowerShell 命令相关的 PowerShell 基础知识。然后，建议您按照章节顺序进行阅读与学习，并按照本书中的 PowerShell 命令示例尝试进行 Office 365 的管理操作。通过阅读本书，最终您将有能力使用 PowerShell 来管理 Office 365。

注意：务必要先在测试环境中进行实践，在没有完全搞清楚命令的真正用途前，不要在生产环境中盲目地使用任何命令。您可以参考在本书的第 1 章中介绍的注册 Office 365 测试账户的基本步骤，来搭建属于自己的测试环境。

命令和输出

为了使读者更容易分辨 PowerShell 的执行命令和命令输出结果，全书命令和输出结果将以如下方式显示：

PS > Get-Help Get-LocalUser -Parameter SID

```
-SID <SecurityIdentifier[]>
    Required?                    false
    Position?                    0
    Accept pipeline input?       true (ByValue, ByPropertyName)
    Parameter set name           SecurityIdentifier
    Aliases                      None
    Dynamic?                     false
```

命令提示符

在管理 Office 365 时，需要创建不同角色的管理员来分担不同的管理任务。本书按照这一原则，创建了不同的用户，并为这些用户分配了不同的管理角色，使他们具有相应的 Office 365 应用和服务的管理权限。全书以命令提示符的方式来区分执行命令的不同管理员，其中主要涉及的用户身份见下表。

提示符	用户及其权限
O365 PS >	admin，租户管理员，也称全局管理员，在一个租户内有最高权限
EXO PS >	exo Admin，具有 Exchange 管理员的权限
SFB PS >	sfb Admin，具有 Skype for Business 管理员的权限
SPO PS >	spo Admin，具有 SharePoint 管理员的权限
USER01 PS >	表示 user 01，仅具有普通用户的权限

示意图

为方便读者理解，书中给出了部分操作过程的示意图。这些图片有可能与您看到的操作界面存在区别。产生区别的原因，一方面是由于 Office 365 的服务处于不断更新的过程中；另一方面，操作界面的呈现效果与您系统使用的语言、版本都有关系。

例如，根据您的系统和语言环境，您看到的开始菜单也许是"开始"，也许是"Start"，对于这种情况，书中则使用中文"开始菜单"表述，不作分别描述。

建议您在学习过程中参照本书的方法，对照您的操作界面进行实际操作。

基本概念

在本书中，经常出现 "租户""用户""订阅"等词，这些词语在微软云系列（包括 Microsoft Azure、Office 365 等）产品中，有特定的含义，请阅读时注意：

租户：租户是包含整个组织的 Azure Active Directory（Azure AD）的实体。一个租户至少包含一个订阅和用户。

用户：用户是只与一个租户（即所属的组织）关联的个人。用户是登录到 Office 365 或 Microsoft Azure 以创建、管理和使用资源的账户。

订阅：用户对于云产品的购买，其实是购买一定期限的使用权，因此称为"订阅"。

软件更名信息

在本书即将问世之际，微软官方将 Office 365 更名为 Microsoft 365。为了方便目前大部分用户的使用习惯，本书沿用 Office 365 这一名称，特此说明。

世纪互联蓝云与世纪互联蓝云研究院

世纪互联蓝云、世纪互联蓝云公司均为上海蓝云网络科技有限公司的简称，是世纪互联成立的全资子公司。由世纪互联蓝云的工程师们组成的"世纪互联蓝云研究院"，致力于在云计算、云应用行业进行技术的推广和研究，并为读者提供更多云计算、云应用等领域新技术的应用技术图书。

目 录

第 1 章 使用 PowerShell 管理 Office 365 的基础知识 ········· 1

1.1 什么是 PowerShell ········· 1
   1.1.1 如何输入 PowerShell 的命令 ········· 1
   1.1.2 如何使用 PowerShell 的帮助系统 ········· 2
1.2 为什么要使用 PowerShell 管理 Office 365 ········· 3
1.3 为什么要注册 Office 365 测试账户，如何完成注册 ········· 4
1.4 使用 PowerShell 管理 Office 365 需要做哪些准备 ········· 7
   1.4.1 如何下载和安装公共组件 ········· 7
   1.4.2 如何解决 AAD Module 的安装问题 ········· 9
1.5 可以用 PowerShell 管理 Office 365 中的所有组件吗 ········· 12
1.6 如何验证 Office 365 的 PowerShell 管理模块已经就绪 ········· 14
1.7 Office 365 的 PowerShell 管理命令有哪些特点 ········· 16

第 2 章 使用 PowerShell 管理 Office 365 的租户 ········· 17

2.1 如何使用 PowerShell 连接 Office 365 的租户 ········· 17
2.2 如何使用 PowerShell 管理 Office 365 的用户 ········· 19
   2.2.1 如何创建和删除用户 ········· 20
   2.2.2 如何查询用户及其属性 ········· 27
   2.2.3 如何修改用户的属性 ········· 33
   2.2.4 如何批量操作用户 ········· 37
2.3 如何使用 PowerShell 管理 Office 365 的角色和组 ········· 42

2.3.1　如何管理 Office 365 的角色 ·· 42

　　　2.3.2　如何管理 Office 365 的安全组 ·· 47

　2.4　如何使用 PowerShell 管理 Office 365 的订阅和许可证 ··································· 54

　　　2.4.1　如何查看订阅及相关信息 ··· 55

　　　2.4.2　如何为用户分配许可证 ·· 58

　　　2.4.3　如何对许可证进行定制化修改 ·· 60

第 3 章　使用 PowerShell 管理 Office 365 的 Exchange Online ················· 63

　3.1　如何使用 PowerShell 连接 Exchange Online ··· 63

　3.2　如何使用 PowerShell 管理 Exchange Online 的收件人 ································ 68

　　　3.2.1　如何管理邮箱 ·· 68

　　　3.2.2　如何管理组 ··· 90

　　　3.2.3　如何管理资源 ·· 104

　　　3.2.4　如何管理联系人 ··· 107

　　　3.2.5　如何管理共享邮箱 ·· 113

　3.3　如何使用 PowerShell 管理 Exchange Online 的权限 ································· 120

　　　3.3.1　如何管理管理员角色 ··· 120

　　　3.3.2　如何管理用户角色 ·· 131

　　　3.3.3　如何管理 Outlook Web App 策略 ·· 138

第 4 章　使用 PowerShell 管理 Office 365 的 Skype for Business Online ·· 143

　4.1　如何使用 PowerShell 连接 Skype for Business ·· 143

　4.2　如何使用 PowerShell 管理 Skype for Business 的用户 ······························ 149

　　　4.2.1　如何查看用户 ·· 149

　　　4.2.2　如何查看用户的属性 ··· 151

　　　4.2.3　如何修改用户的属性 ··· 160

　4.3　如何使用 PowerShell 管理 Skype for Business 的组织 ······························ 163

　　　4.3.1　如何管理"状态隐私模式"属性 ··· 164

　　　4.3.2　如何管理"移动电话通知"属性 ··· 165

　　　4.3.3　如何管理"外部通信"属性 ··· 166

第 5 章　使用 PowerShell 管理 Office 365 的 SharePoint Online……172

5.1　如何使用 PowerShell 连接 SharePoint Online……173
5.2　如何使用 PowerShell 管理 SharePoint Online 的网站……177
5.2.1　如何查看网站与修改网站的属性……177
5.2.2　如何添加与删除网站……181
5.2.3　如何测试与修复网站……185
5.3　如何使用 PowerShell 管理 SharePoint Online 网站中的组……187
5.3.1　如何查询网站中的组……187
5.3.2　如何管理网站中的组……191
5.4　如何使用 PowerShell 管理 SharePoint Online 网站中的用户……193
5.4.1　如何管理外部共享……194
5.4.2　如何管理外部用户……197
5.4.3　如何管理网站中的用户……207
5.5　如何使用 PowerShell 管理 OneDrive for Business 个人网站……217
5.5.1　如何查看个人网站……217
5.5.2　如何制备个人网站……219
5.5.3　如何管理个人网站……220

附录 A　本书所涉及的 PowerShell 知识点……225

附录 B　本书所涉及的 PowerShell for Office 365 命令……234

第 1 章　使用 PowerShell 管理 Office 365 的基础知识

1.1　什么是 PowerShell

PowerShell 是运行在 Windows 系统上的脚本环境和命令行外壳程序。通过使用 PowerShell，用户和系统管理员可以操作.NET Framework 的对象，可以配置管理 Windows 操作系统和其他微软的软件。

1.1.1　如何输入 PowerShell 的命令

我们首先需要知道在哪里可以输入 PowerShell 的命令。如果我们使用的是 Windows 10 操作系统，在系统的开始菜单中就可以找到 PowerShell 的快捷执行方式，如图 1-1 所示。

Windows PowerShell ISE 是 PowerShell 的集成脚本环境，其中包含命令提示和自动补全等便捷特性，而 Windows PowerShell 是 PowerShell 的交互式窗口。在集成脚本环境和交互式窗口中都可以键入 PowerShell 命令。

用鼠标右键单击 Windows PowerShell 菜单项，在弹出的快捷菜单中单击 More 按钮，并在接下来弹出的扩展菜单中单击 Run as Administrator 按钮，如图 1-2 所示，即可使用管理员的权限来运行 PowerShell 的命令。

图 1-1

如果在开始菜单中没有找到 PowerShell 的快捷执行方式，一般可以在以下目录中找到 PowerShell 执行程序：

%SystemRoot%\system32\WindowsPowerShell\v1.0

除了可以在集成脚本环境和交互式窗口中键入 PowerShell 命令并查看输出结果，还可以把 PowerShell 命令写入后缀为.ps1 的文件，然后执行这个脚本文件。

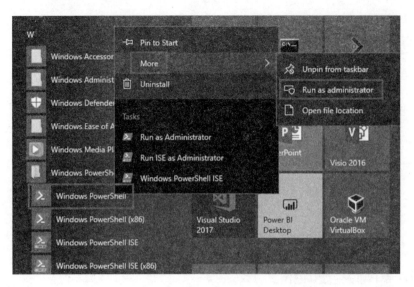

图 1-2

如果用户可以在集成脚本环境或交互式窗口中正常执行 PowerShell 命令，但是，在把 PowerShell 命令保存成 ".ps1" 文件后，执行时却出现错误；或是在连接 Exchange Online 执行 Import-PSSession 命令时出现以下错误：Files cannot be loaded because running scripts is disabled on this system，这大多数是由于系统安全设置级别导致的。

可以尝试使用管理员权限打开 PowerShell 控制台，然后执行以下命令来修改系统安全设置级别，命令如下：

O365 PS > Set-ExecutionPolicy RemoteSigned

1.1.2 如何使用 PowerShell 的帮助系统

PowerShell 采用模块化的组织方式管理微软不同的产品或服务。每个 PowerShell 模块包含数量不等的命令，每条命令又包含多个参数。要想把这些命令全部都记住并灵活使用，确实有点难度。然而，我们可以利用微软在官方网站上为 PowerShell 用户提供的在线帮助手册，随时查阅相关命令及其参数。管理 Office 365 各个 PowerShell 模块的在线帮助手册的网址参见表 1-1。

表 1-1

PowerShell 模块	在线帮助手册地址
MSOnline	https://docs.microsoft.com/en-us/powershell/azure/active-directory/overview?view=azureadps-1.0
Exchange Online	https://docs.microsoft.com/en-us/powershell/exchange/exchange-online/exchange-online-powershell?view=exchange-ps
Skype for Business	https://docs.microsoft.com/en-us/powershell/skype/intro?view=skype-ps
SharePoint Online	https://docs.microsoft.com/en-us/powershell/module/sharepoint-online/?view=sharepoint-ps

在集成脚本环境和命令行交互式窗口中使用 Get-Command 和 Get-Help 命令可以获取帮助信息。PowerShell 帮助命令的具体使用方法请参考本书的附录 A。

1.2 为什么要使用 PowerShell 管理 Office 365

通过使用 PowerShell，用户可以在统一的操作界面上管理 Windows 操作系统和微软其他的应用软件。例如，操作系统 Windows Server 2012 中的大部分功能。都可以通过使用 PowerShell 或使用基于 PowerShell 的图形化工具来进行配置。Office 365 管理界面上的重要功能也都可以通过使用 PowerShell 来进行操作。有些在图形界面上无法完成的工作，借助 PowerShell 也可以轻松完成。目前，用于管理 Exchange Online 的 PowerShell 命令已经达到 729 个，管理 Skype for Business Online 的命令 393 个，管理 SharePoint Online 的命令 186 个，而且，这些 PowerShell 管理命令的数量还在增加。作为 Office 365 的管理员，能够掌握并灵活使用这些命令将是大势所趋。

使用 PowerShell 的循环操作来管理 Office 365 可以大幅度提高管理的效率，快捷方便地达到管理的目标，将管理员从简单枯燥的重复性劳动中解放出来。

使用 PowerShell 的管道操作，可以方便快捷地对结果进行过滤，迅速找到我们关注的部分，还可以对结果的输出格式进行调整，对需要输入的内容使用 PowerShell 进行再加工和控制，提高操作的灵活程度。

为了让读者可以从熟悉的图形界面入手，并通过比较图形管理界面和 PowerShell 命令的异同来感受使用 PowerShell 命令的种种好处，我们在书中一般先简要介绍各项功能在 Office 365 管理中心的图形界面上的操作方法。

要访问Office 365的管理中心，首先需要在浏览器地址栏中键入以下地址：

https://login.partner.microsoftonline.cn

在页面弹出的"登录"对话框中，如图1-3所示，键入租户管理员的账户[①]和密码后，就可以登录Office 365的门户网站了。

登录后，单击页面左上角的九宫格形状的 app launcher 按钮，并在弹出的扩展菜单中单击 Admin 按钮，来打开 Office 365 的管理中心，如图 1-4 所示。

图 1-3

图 1-4

---

① 本书软件界面的"帐户""帐号""帐单"，分别应为"账户""账号""账单"，本书正文中使用"账户""账号""账单"。

我们也可以直接在浏览器中键入以下网址来打开Office 365的管理中心：
https://portal.partner.microsoftonline.cn/AdminPortal

另外，如果要改变语言设置，在登录Office 365的管理中心后，可以单击页面右上角齿轮形状的设置按钮，然后，在弹出的菜单中单击View all按钮，如图1-5所示；并在弹出的扩展菜单中的"语言"下拉列表框中选择"中文（中国）"，单击"保存"按钮，如图1-6所示。

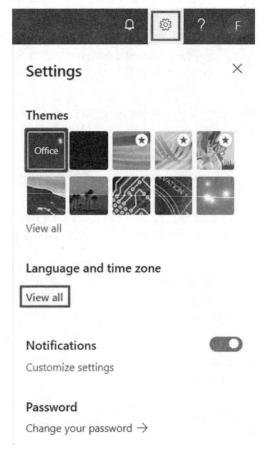

图1-5　　　　　　　　　　　　　　图1-6

1.3　为什么要注册Office 365测试账户，如何完成注册

作为系统管理人员，我们要坚持的铁律之一是不能在生产环境中进行未经测试的操作，以及运行未经测试的脚本命令。这样会在一定程度上给我们的学习过程造成困难。为了解决这个问题，我们可以注册一个免费试用30天的Office 365试用账号，参考步骤如下。

（1）在浏览器中键入以下地址：
https://products.office.com/zh-CN/
根据页面的提示单击"商业版"按钮。

（2）在接下来的"为你找到合适的解决方案"页面的左侧"满足中国用户的需求"项下，单击"查看Microsoft 365中国版订阅计划"链接，如图1-7所示。

图 1-7

（3）在下一个"比较面向中小企业的中国版本计划和价格"页面中的"Microsoft 365 商业标准版"[①]项目下，单击"免费试用 1 个月"链接，如图 1-8 所示。

图 1-8

---

① 以前的"Office 365 商业高级版"现更名为"Microsoft 365 商业标准版"。

（4）填写姓名、邮箱及公司名称等信息，并单击"下一页"按钮，如图1-9所示。

图1-9

（5）设置用户ID和密码，然后单击"创建我的账户"按钮，如图1-10所示。

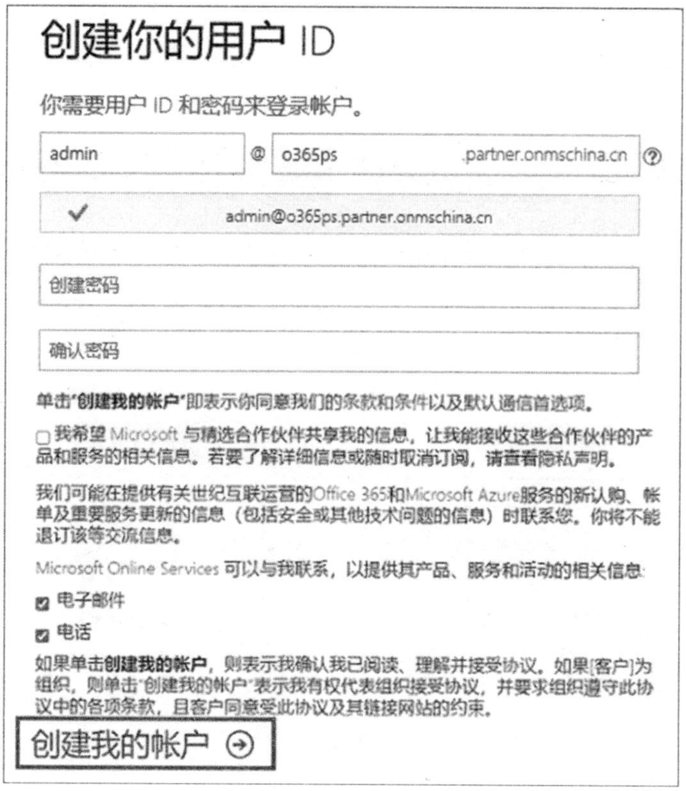

图1-10

（6）建议保存以下页面中的重要信息，以便日后可以使用，并单击"您已准备就绪…"按钮，如图 1-11 所示。

图 1-11

> **注意**
>
> 虽然 Office 365 已经更名为 Microsoft 365，且"Office 365 商业高级版"也变更为"Microsoft 365 商业标准版"。但是，为方便多数用户的使用习惯，本书仍将沿用"Office 365"和"Office 365 商业高级版"的称谓。

1.4　使用 PowerShell 管理 Office 365 需要做哪些准备

使用 PowerShell 管理 Office 365 的准备工作包括相关组件的下载和安装，以及解决安装过程中出现的问题。

PowerShell 对以下操作系统都支持：Windows 10、Windows 8.1、Windows 8、Windows 7 Service Pack 1 (SP1)、Windows Server 2016、Windows Server 2012 R2、Windows Server 2012、Windows Server 2008 R2 SP1。建议使用 64 位版本的操作系统，因为 32 位版本的 Microsoft Azure Active Directory Module for Windows PowerShell（简称 AAD Module）在 2014 年年底就已经不再更新了。

1.4.1　如何下载和安装公共组件

1. 安装 Microsoft Online Services Sign-in Assistant

Microsoft Online Services Sign-in Assistant 安装包可以从下面的地址下载：
http://www.microsoft.com/en-us/download/details.aspx?id=41950
下载之前，请确认操作系统的语言版本，如图 1-12 所示。建议使用 64 位版本的安装包。

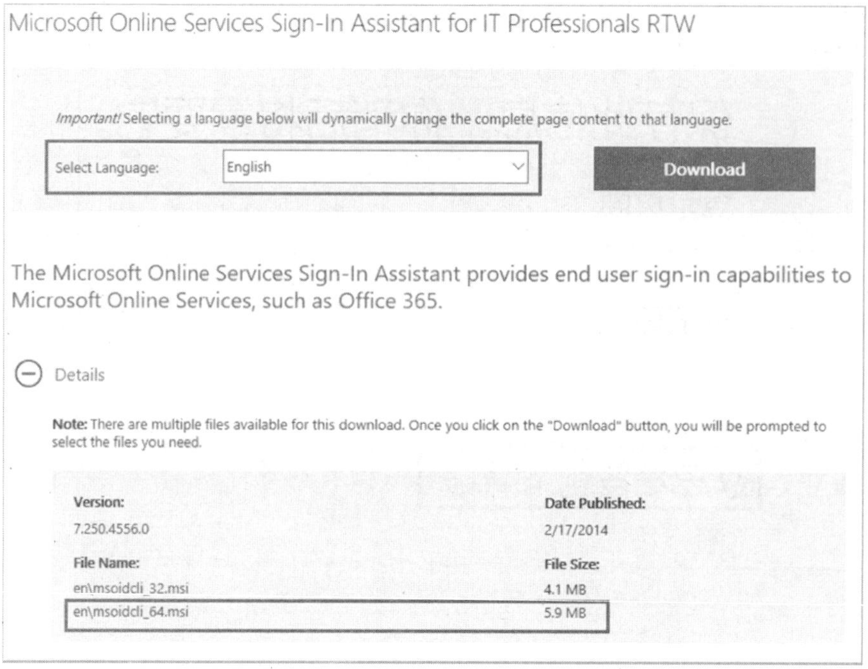

图 1-12

下载完成之后，双击安装包，选择"同意许可证以及隐私条款"复选框，并单击"下一步"按钮，就可以开始安装了。

2. 安装 Microsoft Azure Active Directory Module for Windows PowerShell

接下来安装 Microsoft Azure Active Directory Module for Windows PowerShell（简称 AAD Module）。这个模块可以在控制台执行 PowerShell 命令进行安装。前提是需要用管理员权限打开 PowerShell 的控制台。操作步骤如下：

（1）用鼠标右键单击 Windows PowerShell 的菜单项，在弹出的快捷菜单中单击 Run as Administrator 按钮。

注意

这时打开的控制台在窗口标题栏有 Administrator 字样，即 Administrator: Windows PowerShell，而使用普通用户权限打开的控制台在窗口标题栏仅有 Windows PowerShell。

（2）执行下面的 Install-Module 命令，同意安装 NuGet provider，并同意从 PSGallery 安装模块，命令和输出结果如下：

O365 PS > Install-Module MSOnline

```
NuGet provider is required to continue
PowerShellGet requires NuGet provider version '2.8.5.201' or newer to interact
with NuGet-based repositories. The NuGet provider must be available in 'C:\Program
Files\PackageManagement\ProviderAssemblies' or
```

```
'C:\Users\Administrator\AppData\Local\PackageManagement\ProviderAssemblies'.
You can also install the NuGet provider by running 'Install-PackageProvider -Name
NuGet -MinimumVersion 2.8.5.201 -Force'. Do you want PowerShellGet to install and
import the NuGet provider now?
  [Y] Yes  [N] No  [S] Suspend  [?] Help (default is "Y"): Y

Untrusted repository
You are installing the modules from an untrusted repository. If you trust this
repository, change its InstallationPolicy value by running the Set-PSRepository
cmdlet. Are you sure you want to install the modules from 'PSGallery'?
  [Y] Yes  [A] Yes to All  [N] No  [L] No to All  [S] Suspend  [?] Help (default
is "Y"): Y
```

执行完成后，就可以关闭这个 PowerShell 的控制台窗口了。

1.4.2 如何解决 AAD Module 的安装问题

使用 Install-Module 命令安装 AAD Module 时，有时会出现下面的报错信息。命令和输出结果如下：

O365 PS > Install-Module MSOnline

```
Install-Module: The term 'Install-Module' is not recognized as the name of a
cmdlet, function, script file, or operable program. Check the spelling of the name,
or if a path was included, verify that the path is correct and try again.
At line:1 char:1
+ Install-Module MSOnline
+ ~~~~~~~~~~~~~~
    + CategoryInfo          : ObjectNotFound: (Install-Module:String) [],
CommandNotFoundException
    + FullyQualifiedErrorId : CommandNotFoundException
```

这个错误是由于本机的 PowerShell 版本过低导致的，我们可以使用下面的命令查看一下本机安装的 PowerShell 版本。命令和输出结果如下：

O365 PS > $PSVersionTable

```
Name                           Value
----                           -----
PSVersion                      4.0
WSManStackVersion              3.0
SerializationVersion           1.1.0.1
CLRVersion                     4.0.30319.42000
BuildVersion                   6.3.9600.18968
PSCompatibleVersions           {1.0, 2.0, 3.0, 4.0}
PSRemotingProtocolVersion      2.2
```

> **注意**

这里 PSVersion 的值是 4.0，这个值也可能是 3.0，都表明 PowerShell 版本较低。在这些低版本的 PowerShell 中，缺少 PowerShellGet，所以无法使用 Install-Module 这条命令。PowerShellGet 是 PowerShell 的安装包管理工具，是 Windows 组件 OneGet 的外壳程序，主要是为了简化 PowerShell 模块的包管理工作而设计的。

以下两种方法可以解决 PowerShell 版本过低的问题。

1. 升级低版本的 PowerShell 到 5.1 版本

下载 Windows Management Framework (WFM) 5.1 版本的安装包并进行安装，将目前低版本的 PowerShell 升级成 5.1 版本（这里仅考虑 64 位版本的操作系统）。不同操作系统所需的安装包及下载地址请参见表 1-2。

表 1-2

操作系统	安装包	下载地址
Windows 7 SP1	Win7AndW2K8R2-KB3191566-x64.ZIP	http://download.microsoft.com/download/6/F/5/6F5FF66C-6775-42B0-86C4-47D41F2DA187/Win7AndW2K8R2-KB3191566-x64.zip
Windows Server 2008 R2 SP1	Win7AndW2K8R2-KB3191566-x64.ZIP	http://download.microsoft.com/download/6/F/5/6F5FF66C-6775-42B0-86C4-47D41F2DA187/Win7AndW2K8R2-KB3191566-x64.zip
Windows 8.1	Win8.1AndW2K12R2-KB3191564-x64.msu	http://download.microsoft.com/download/6/F/5/6F5FF66C-6775-42B0-86C4-47D41F2DA187/Win8.1AndW2K12R2-KB3191564-x64.msu
Windows Server 2012	W2K12-KB3191565-x64.msu	http://download.microsoft.com/download/6/F/5/6F5FF66C-6775-42B0-86C4-47D41F2DA187/W2K12-KB3191565-x64.msu
Windows Server 2012 R2	Win8.1AndW2K12R2-KB3191564-x64.msu	http://download.microsoft.com/download/6/F/5/6F5FF66C-6775-42B0-86C4-47D41F2DA187/Win8.1AndW2K12R2-KB3191564-x64.msu

如果操作系统是 Windows 7 SP1 或 Windows Server 2008 R2 SP1，将下载的 ZIP 文件解压缩后会得到一个 MSU 安装文件和一个 PowerShell 的安装脚本，建议把它们放在同一文件夹下；接着用管理员权限打开 PowerShell 的控制台，切换到刚才存放安装文件和脚本的文件夹下，执行 Install-Wmf5.1.ps1 这个安装脚本，并按照屏幕提示完成剩下的安装工作就可以了。

如果操作系统是 Windows 8.1、Windows Server 2012 或 Windows Server 2012 R2，安装过程会更加简单一些。只需要进入下载后存放安装文件的目录中，双击并运行安装文件，按照屏幕提示完成安装工作即可。

如果操作系统的版本是 Windows 7 SP1 或 Windows Server 2008 R2 SP1，在升级 PowerShell 的版本之前，需要先安装.NET Framework 4.5.2。

（1）从下面这个地址下载并安装.NETFramework 4.5.2：

https://www.microsoft.com/en-ca/download/details.aspx?id=42642

下载之前按照操作系统的语言选择安装包的语言，然后单击"下载"按钮开始下载，如图 1-13 所示。

第 1 章　使用 PowerShell 管理 Office 365 的基础知识

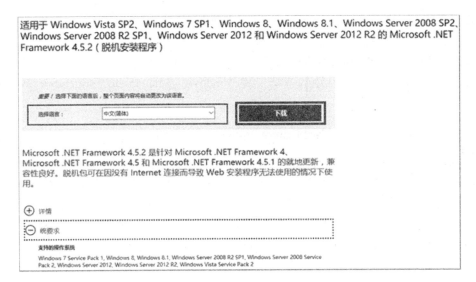

图 1-13

（2）如果需要先安装 .NET Framework 4.5.2，在安装之前，还要记得提前卸载已经安装的 WMF 的 3.0 版本，否则会导致 PSModulePath 这个 Windows 的环境变量丢失，进而导致其他应用程序出错。我们可以提前备份系统环境变量当中的 PSModulePath，并在 WMF 5.1 安装完成之后，再把之前备份的环境变量路径还原回去。备份系统环境变量的具体操作步骤如下：

打开控制面板 Control Panel | All Control Panel Items，单击 System 按钮打开系统对话框，单击左侧导航栏中的 Advanced system settings 按钮打开系统属性对话框，在 Advanced 标签中单击对话框底部的 Environment Varialbes 按钮，打开环境变量对话框，备份 System Varialbes（系统变量）项下的 PSModulePath 属性值，如图 1-14 所示。

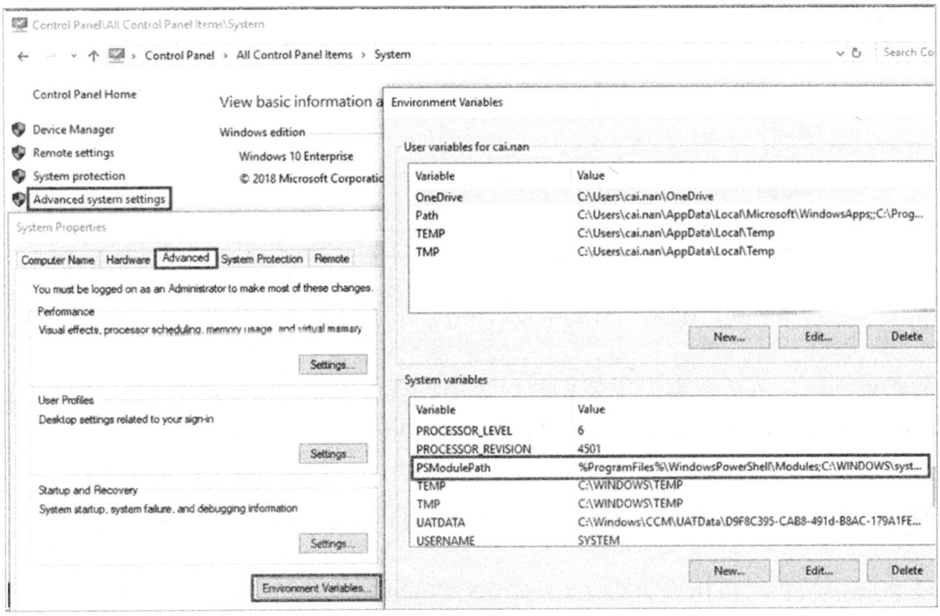

图 1-14

卸载 WMF 3.0 的步骤如下：首先，进入"控制面板"，找到"应用程序"，在"应用程序和功能"项下查看已经安装的更新，并将之前安装的这个更新卸载。一般情况下，在 Windows 7 SP1 和 Windows Server 2008 R2 SP1 系统中这个 Windows 更新的名称是 KB2506143。

接下来，就可以按照之前介绍的方法安装 Windows Management Framework (WFM) 的 5.1 版本了。

2. 安装 PackageManagement PowerShell Modules Preview

这种解决方法是在暂时不方便对现有低版本的 PowerShell 进行升级时使用的，思路是通过安装 PowerShell 的安装包管理工具使 Active Directory Module for Windows PowerShell 模块可以正常进行安装。

先从以下地址下载 PackageManagement PowerShell Modules Preview：
https://www.microsoft.com/en-us/download/details.aspx?id=51451

如果操作系统是 64 位版本的，请下载 x64 版本的安装包，如图 1-15 所示。

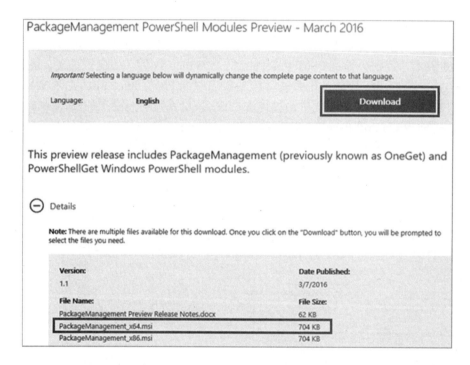

图 1-15

这个安装包只有英文版本。安装正常结束后，就可以使用 Install-Module MSOnline 这个 PowerShell 命令来安装 Active Directory Module for Windows PowerShell 了。

1.5 可以用 PowerShell 管理 Office 365 中的所有组件吗

使用 PowerShell 可以用来管理 Office 365 中的所有组件。为了管理 Skype for Business Online 和 SharePoint Online，需要进行以下准备工作。

1. 安装用于管理 Skype for Business Online 的组件

以下操作系统的 64 位版本均支持使用 PowerShell 来管理 Skype for Business Online：Windows 7、Windows Server 2008、Windows 8、Windows 8.1、Windows 10、Windows Server 2012 和 Windows Server 2012 R2。

在确认过操作系统的版本后，就可以开始安装适用于 Skype for Business Online 的 Windows PowerShell 管理模块了。下载地址是：

https://www.microsoft.com/zh-cn/download/details.aspx?id=39366

在安装包的下载页面中选择正确的语言版本后，单击"下载"按钮开始下载，如图 1-16 所示。

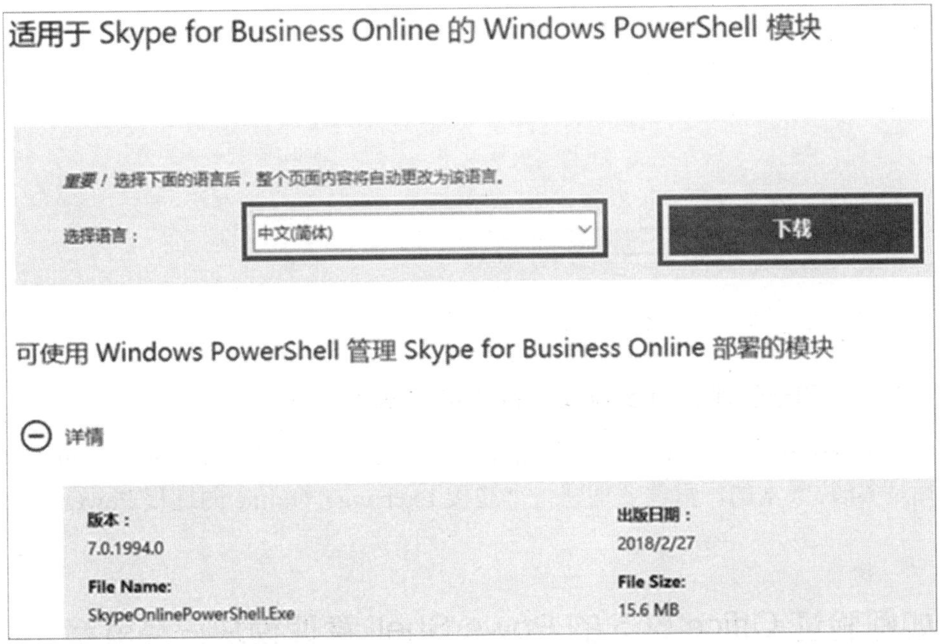

图 1-16

下载完成后，依照屏幕提示进行安装。

2. 安装用于管理 SharePoint Online 的组件

使用 PowerShell 来管理 SharePoint Online 不要求操作系统是 64 位版本的，这一点和 Skype for Business Online 不同。类似的是，需要安装 SharePoint Online Management Shell 模块，下载地址如下：

https://www.microsoft.com/zh-cn/download/details.aspx?id=35588

如果操作系统版本是 64 位的，可以下载并安装 64 位 SharePoint 管理工具，如图 1-17 所示。

选择好语言版本后，就可以下载和安装了。

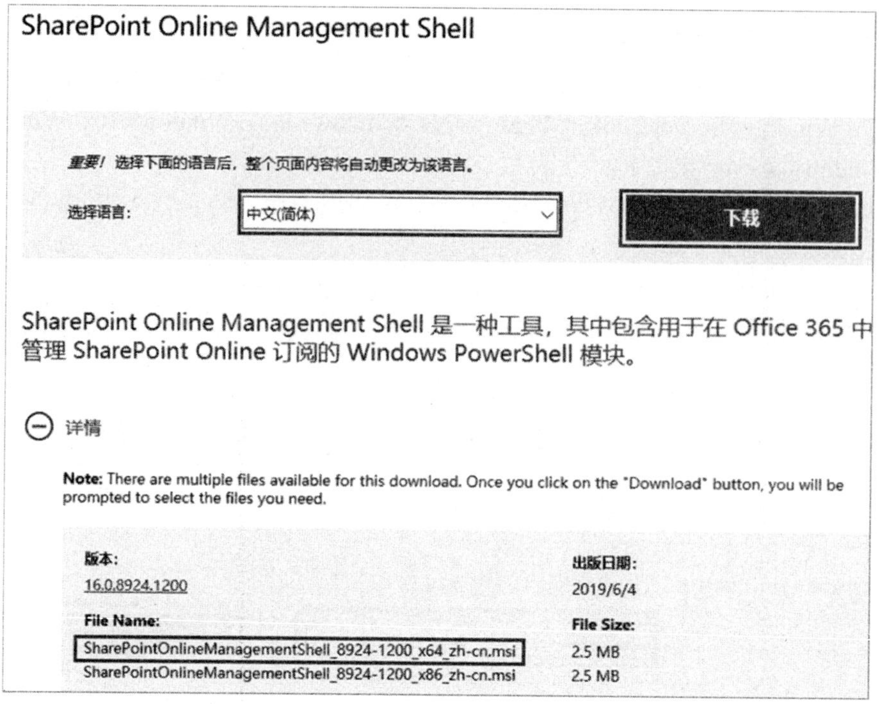

图 1-17

3. 无须安装用于管理 Exchange Online 的额外组件

使用 PowerShell 来管理 Exchange Online，不需要下载额外的安装模块，也不要求操作系统必须是 64 位版本的，只需要创建一个连接 Exchange Online 的远程 PowerShell 会话就可以了。

1.6 如何验证 Office 365 的 PowerShell 管理模块已经就绪

我们可以分别加载 MSOnline 模块、Skype for Business Online 模块和 SharePoint Online 模块，并通过查询相应管理模块的名称和版本来验证上述模块是否已经正常安装。

1. 加载 MSOnline 模块

（1）查看系统默认已经加载的模块。具体步骤是，打开一个新的 PowerShell 控制台窗口，键入 Get-Module 命令，命令和输出结果如下：

O365 PS > Get-Module | Select-Object ModuleType,Version,Name

```
ModuleType Version Name
---------- ------- ----
  Manifest 3.1.0.0 Microsoft.PowerShell.Management
  Manifest 3.1.0.0 Microsoft.PowerShell.Utility
    Script 1.2     PSReadline
```

（2）使用以下命令加载 Active Directory Module for Windows PowerShell 模块：

O365 PS > Import-Module MSOnline

（3）再运行一次 Get-Module 命令，查看 MSOnline 模块的加载情况。命令和输出结果如下：

O365 PS > Get-Module | Select-Object ModuleType,Version,Name

```
ModuleType Version   Name
---------- -------   ----
  Manifest 3.1.0.0   Microsoft.PowerShell.Management
  Manifest 3.1.0.0   Microsoft.PowerShell.Utility
  Manifest 1.1.166.0 MSOnline
  Script   1.2       PSReadline
```

2. 加载 Skype for Business Online 模块

使用下面的命令加载 Skype for Business Online 模块，并检查模块的加载情况。命令和输出结果如下：

O365 PS > Import-Module LyncOnlineConnector
O365 PS > Get-Module | Select-Object ModuleType,Version,Name

```
ModuleType Version   Name
---------- -------   ----
  Manifest 7.0.0.0   LyncOnlineConnector
  Manifest 3.1.0.0   Microsoft.PowerShell.Management
  Manifest 3.1.0.0   Microsoft.PowerShell.Utility
  Manifest 3.0.0.0   Microsoft.WSMan.Management
  Manifest 1.1.166.0 MSOnline
  Script   1.2       PSReadline
  Script   7.0.0.0   SkypeOnlineConnector
```

3. 加载 SharePoint Online 模块

使用下面的命令加载 SharePoint Online 模块，并检查模块的加载情况。命令和输出结果如下：

O365 PS > Import-Module Microsoft.Online.SharePoint.PowerShell -DisableNameCheck
O365 PS > Get-Module | Select-Object ModuleType,Version,Name

```
ModuleType Version     Name
---------- -------     ----
  Manifest 7.0.0.0     LyncOnlineConnector
  Binary   16.0.7521.0 Microsoft.Online.SharePoint.PowerShell
  Manifest 3.1.0.0     Microsoft.PowerShell.Management
  Manifest 3.1.0.0     Microsoft.PowerShell.Utility
```

```
Manifest   3.0.0.0       Microsoft.WSMan.Management
Manifest   1.1.166.0     MSOnline
 Script    1.2           PSReadline
 Script    7.0.0.0       SkypeOnlineConnector
```

1.7　Office 365 的 PowerShell 管理命令有哪些特点

虽然 PowerShell 命令的绝对数量很多，但是这些命令还是有特点的，它们都有着相似的形式，即"动词-<模块名称>名词"。注意，动词后面还有个连字符。

以下这些动词经常出现：Get、Set、Add、New、Enable、Disable、Start、Stop、Invoke、Remove，其中有些动词还是成对出现的。模块的名称包括 Msol、Cs、SPO 等，名词包括 User、Group、License 等，这里就不逐一列举了，当我们用到的时候再进行解释。通过归类和分组，可以帮助我们记住大多数常用的命令。

第 2 章 使用 PowerShell 管理 Office 365 的租户

2.1 如何使用 PowerShell 连接 Office 365 的租户

本章将使用租户的全局管理员 Admin 的用户凭据来进行相关的管理操作。

1. 连接租户的具体步骤

按照第 1 章的相关内容进行准备后，我们已经有了可以正常连接 Office 365 租户的系统环境了。接下来，我们可以按照 1.1.1 节介绍的内容，打开 PowerShell 的交互式窗口或者 PowerShell 的集成脚本环境，输入以下命令，来尝试连接这个租户，命令如下：

O365 PS > Import-Module MSOnline
O365 PS > Connect-MsolService

执行上述命令后，将会弹出以下 Office 365 的登录对话框，如图 2-1 所示。

图 2-1

使用我们之前注册的 Office 365 试用版的管理员账户和密码尝试进行登录，会收到以下报错提示，如图 2-2 所示。

图 2-2

这个报错是由于我们安装了最新版本的 Microsoft Azure Active Directory Module for Windows PowerShell。而在使用这个新版本的 AAD Module 连接中国版 Office 365 时，需要添加参数 AzureEnvironment。命令和输出结果如下：

O365 PS > Import-Module MSOnline

O365 PS > $cred = Get-Credential

```
cmdlet Get-Credential at command pipeline position 1
Supply values for the following parameters:
Credential
```

O365 PS > Connect-MsolService -Credential $cred -AzureEnvironment AzureChinaCloud

从 PowerShell 的 3.0 版本开始，就不再需要使用 Import-Module MSOnline 这条命令了，因为当执行 Connect-MsolService 命令时会自动加载 MSOnline 模块。

PowerShell 命令 Get-Credential 会打开下面这个对话框，如图 2-3 所示。

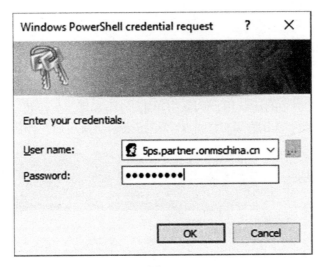

图 2-3

键入 Office 365 试用版的管理员账户和密码，这个用户凭据会被保存在 $cred 这个变量中，并通过 Credential 这个参数传递给 Connect-MsolService 命令。

我们可以把上述这三条命令保存起来，或者存储成 PowerShell 的 .ps1 脚本，这样在每次需要进行登录操作时，直接运行就可以了。

2. 查询本租户的域名和相关信息

运行下面的命令来查询本租户的域名以及相关的信息，从输出的结果我们可以得到这个租户的域名、域名的当前状态和认证的方式等信息。命令和输出结果如下：

O365 PS > Get-MsolDomain

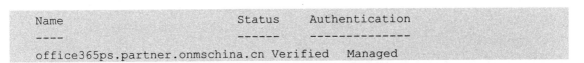

```
Name                              Status    Authentication
----                              ------    --------------
office365ps.partner.onmschina.cn  Verified  Managed
```

至此，我们已经成功连接我们的测试租户了。

2.2 如何使用 PowerShell 管理 Office 365 的用户

对于系统管理员来说，日常最频繁的工作应该就是用户的维护工作了。在 Office 365 的管理中心中，用户管理的功能模块也被放在触手可及的位置上，如图 2-4 所示。

图 2-4

2.2.1 如何创建和删除用户

1. 用户创建命令的必选参数

在 Office 365 管理中心的活跃用户列表中单击"+添加用户"按钮，在页面右侧弹出的对话框中键入用户的信息，单击底部的"添加"按钮来创建新用户，如图 2-5 所示。

图 2-5

使用命令 New-MsolUser 可以实现同样的功能。我们先来看看这条命令最基础的用法。命令和输出结果如下：

```
O365 PS > New-MsolUser `
>> -UserPrincipalName "user02@office365ps.partner.onmschina.cn" `
>> -DisplayName "user 02"

Password  UserPrincipalName                          DisplayName  isLicensed
--------  -----------------                          -----------  ----------
A356ez39  user02@office365ps.partner.onmschina.cn    user 02      False
```

注意

（1）参数 UserPrincipalName 和 DisplayName 是必选项，换句话说，如果缺少这两个参数当中的任何一个，这个命令都不会成功完成。

（2）用户名和显示名称是两个不同的概念。用户名在租户中是唯一的，而不同的两个用户可以有相同的显示名称。

（3）系统自动为该账户设置了密码，并将通知创建这个账户的系统管理员。

（4）属性 isLicensed 是空值意味着使用上面的命令暂时不会为用户分配任何许可证，对应着新建用户对话框中的"产品许可证"选项下的"创建用户但不分配产品许可证"选项。

2. 用户创建命令的可选参数

在活跃用户的列表中添加新用户时无法设置默认语言，也不能禁用密码复杂度策略；而使用 PowerShell 却可以实现这些特殊的需求。

（1）通过 PreferredLanguage 参数设置用户的默认语言。

即使是使用中国版 Office 365，新创建用户的默认语言也会被设置为英文。例如，公司里的用户 user 04 的母语是中文，而另一个用户 user 05 的母语是英语。

要想在 user 04 登录前就指定其默认语言为中文，而保留 user 05 的语言是美国英语这个默认的设置，可以使用下面的 PowerShell 命令。命令和输出结果如下：

```
O365 PS > New-MsolUser `
>> -UserPrincipalName "user04@office365ps.partner. onmschina.cn" `
>> -DisplayName "user 04" -PreferredLanguage "ZH-CN"

Password  UserPrincipalName                          DisplayName  isLicensed
--------  -----------------                          -----------  ----------
Von34300  user04@office365ps.partner.onmschina.cn    user 04      False
```

```
O365 PS > New-MsolUser `
>> -UserPrincipalName "user05@office365ps.partner. onmschina.cn" `
>> -DisplayName "user 05" -PreferredLanguage "EN-US"
```

```
Password   UserPrincipalName                           DisplayName isLicensed
--------   -----------------                           ----------- ----------
y018tQ12   user05@office365ps.partner.onmschina.cn     user 05     False
```

使用以上 PowerShell 命令创建用户，用户 user 04 登录时看到的界面是简体中文的，而 user 05 看到的是默认的英文界面。

（2）通过 PasswordNeverExpires 参数设置密码永不过期。

通过添加用户对话框创建的新用户，默认设置了密码过期的属性，即到了规定的时间会强制用户修改密码。而使用 New-MsolUser 命令创建新用户时，通过将 PasswordNeverExpires 参数的值设置为 $true（该参数的默认值是 $false），可以使新创建用户的密码永不过期。

如果我们希望某个新创建用户的密码不会过期，一般会在创建用户的同时预先设定好该用户的密码。这就需要在使用 New-MsolUser 命令的时候，同时将 ForceChangePassword 参数的值设置为 $false（该参数的默认值是 $true）。这样，用户在首次登录时，系统就不会提示其更改密码，而可以直接使用管理员预设的密码。命令和输出结果如下：

O365 PS > New-MsolUser `
>> -UserPrincipalName "user06@office365ps.partner. onmschina.cn" `
>> -DisplayName "user 06" -PasswordNeverExpires $true `
>> -ForceChangePassword $false -Password "n2583P3c"

```
Password   UserPrincipalName                           DisplayName isLicensed
--------   -----------------                           ----------- ----------
n2583P3c   user06@office365ps.partner.onmschina.cn     user 06     False
```

（3）通过 StrongPasswordRequired 参数禁用密码的复杂度策略。

在默认情况下，在设置密码或修改密码时需要达到密码复杂度要求的规定。这个规定包括以下内容：密码包括 8 到 16 个字符，且存在大写字母、小写字母、数字和特殊符号当中任意三种的组合。但是在某些特殊情况下，系统管理员可能会希望某些用户的密码尽可能简单，通过修改 StrongPasswordRequired 这个参数的默认值可以达到禁用密码复杂度的需求。命令和输出结果如下：

O365 PS > New-MsolUser `
>> -UserPrincipalName "user07@office365ps.partner.onmschina. cn" `
>> -DisplayName "user 07" -StrongPasswordRequired $false `
>> -ForceChangePassword $false -PreferredLanguage "ZH-CN"

```
Password   UserPrincipalName                           DisplayName isLicensed
--------   -----------------                           ----------- ----------
abV932a8   user07@office365ps.partner.onmschina.cn     user 07     False
```

在成功创建用户 user 07 后，我们用系统为该用户设定的账户凭据登录 Office 365。在键入用户名和密码后，成功登录；接着，单击右上角齿轮形状的"设置"按钮，在弹出的扩展菜单中单击"更改密码"按钮，更改该账户的密码，如图 2-6 所示。

图 2-6

由于我们已经禁用了该用户的密码复杂度策略,在随后弹出的密码变更页面中修改密码时,我们尝试使用 officeps 这个不符合复杂度的密码来进行测试,但是,结果出乎意料,系统依然提示所修改的密码没有满足复杂度的要求,如图 2-7 所示。

另一种类似的情形是,如果创建用户时没有把参数 ForceChangePassword 设置为$False,则用户首次登录时仍会被要求进行密码修改。如果用户修改的密码不符合密码复杂度要求,虽然禁用了密码复杂度策略,也是无法成功修改密码的,如图 2-8 所示。

图 2-7　　　　　　　　　　　　图 2-8

综合上面的几种情况，如果管理员没有在创建用户时手工指定简单的密码或没有要求用户在首次登录时修改密码，那么用户要想尝试把系统设定的复杂密码修改为简单的密码，或在首次登录时尝试设定简单的密码，都是不会成功的。

用下面的方法可以在创建用户时直接为用户设置不满足复杂度要求的密码。命令和输出结果如下：

O365 PS > New-MsolUser `
>> -UserPrincipalName "user09@office365ps.partner.onmschina.cn" `
>> -DisplayName "user 09" -StrongPasswordRequired $false `
>> -ForceChangePassword $false -Password "officeps"

```
Password   UserPrincipalName                          DisplayName  isLicensed
--------   -----------------                          -----------  ----------
officeps   user09@office365ps.partner.onmschina.cn    user 09      False
```

在以上命令中，虽然用户 user 09 的密码并不满足密码复杂度的要求，但是密码可以被设置，且可以正常登录。

注意，虽然我们禁用了密码复杂度策略，但是所设定密码的长度必须是 8 位以上，否则就会报错。命令和输出结果如下：

O365 PS > New-MsolUser `
>> -UserPrincipalName "user10@office365ps.partner.onmschina.cn" `
>> -DisplayName "user 10" -StrongPasswordRequired $false `
>> -ForceChangePassword $false -Password "office"

```
New-MsolUser : You must choose a strong password that contains 8 to 16 characters,
a combination of letters, and at least one number or symbol. Choose another password
and try again.
At line:1 char:1
+ New-MsolUser -UserPrincipalName "user10@office365ps.partner.onmschina ...
+ ~~~~~~~~~~~~~~~~~~~~~~~~~~~~~~~~~~~~~~~~~~~~~~~~~~~~~~~~~
    + CategoryInfo          : OperationStopped: (:) [New-MsolUser],
MicrosoftOnlineException
    + FullyQualifiedErrorId :
Microsoft.Online.Administration.Automation.InvalidPasswordWeakException,Micro
soft.Online.Administration.Automation.NewUser
```

3. 用户删除命令

（1）在 Office 365 管理中心左侧的导航栏中单击"用户"按钮，并在弹出的扩展菜单中单击"活跃用户"按钮，然后在右侧详细内容窗格的活跃用户列表中，选择要删除的用户显示姓名左侧的复选框，如图 2-9 所示。

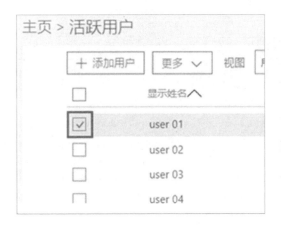

图 2-9

接着,在页面右侧弹出的该用户的属性对话框中,单击"删除用户"按钮,如图 2-10 所示。确认删除操作后,即可删除该用户。

图 2-10

使用 PowerShell 的 Remove-MsolUser 命令配合 ObjectId 或 UserPrincipalName 参数可以删除 Office 365 的用户。命令和输出结果如下:

O365 PS > Remove-MsolUser `
\>\> -UserPrincipalName "user19@office365ps.partner. onmschina.cn"

```
Confirm
Continue with this operation?
[Y] Yes  [N] No  [S] Suspend  [?] Help (default is "Y"): Y
```

参数 UserPrincipalName 只接受完整的用户标识符,如果只提供部分用户标识,即使可以精确匹配到唯一的用户,删除工作也无法成功完成。命令和输出结果如下:

O365 PS > Remove-MsolUser -UserPrincipalName "user20"

```
Confirm
Continue with this operation?
[Y] Yes  [N] No  [S] Suspend  [?] Help (default is "Y"): Y
Remove-MsolUser : User Not Found.  User: user20.
At line:1 char:1
+ Remove-MsolUser -UserPrincipalName "user20"
+ ~~~~~~~~~~~~~~~~~~~~~~~~~~~~~~~~~~~~~~~~~~~
    + CategoryInfo          : OperationStopped: (:) [Remove-MsolUser],
MicrosoftOnlineException
    + FullyQualifiedErrorId :
Microsoft.Online.Administration.Automation.UserNotFoundException,Microsoft.On
line.Administration.Automation.RemoveUser
```

使用下面的 PowerShell 命令，即使只提供部分用户标识，删除工作也能顺利完成。

注意：如果字符串可以匹配多个用户，所有这些符合匹配条件的用户都会被依次删除。命令和输出结果如下：

O365 PS > Get-MsolUser -SearchString "user20" | Remove-MsolUser

```
Confirm
Continue with this operation?
[Y] Yes  [N] No  [S] Suspend  [?] Help (default is "Y"): Y
```

（2）在 Office 365 管理中心左侧的导航栏中单击"用户"按钮，在弹出的扩展菜单中单击"已删除的用户"按钮，即可在右侧的已删除的用户列表中查看已经被删除的用户。

使用 Get-MsolUser 命令并配合 ReturnDeletedUsers 参数，便可以查看已删除的用户。命令和输出结果如下：

O365 PS > Get-MsolUser -ReturnDeletedUsers |
\>\> Select-Object DisplayName,SignInName, SoftDeletionTimestamp

```
DisplayName     SignInName                                    SoftDeletionTimestamp
-----------     ----------                                    ---------------------
user 19         user19@office365ps.partner.onmschina.cn       5/14/2019
10:46:40 AM
user 20         user20@office365ps.partner.onmschina.cn       5/14/2019
10:55:11 AM
```

（3）如果被删除的用户仍处于 30 天的系统保留期内，则可以在已删除的用户列表中选择已删除用户显示姓名左侧的复选框，接下来，在页面右侧弹出的该已删除用户的属性对话框中单击"还原"按钮，来完成删除用户的还原操作。

使用 Restore-MsolUser 命令也可以还原被删除的用户。注意，参数 UserPrincipalName 只接受完整的用户登录名。命令和输出结果如下：

O365 PS > Restore-MsolUser `
\>\> -UserPrincipalName "user20@office365ps.partner.onmschina.cn"

```
UserPrincipalName                         DisplayName isLicensed
-----------------                         ----------- ----------
user20@office365ps.partner.onmschina.cn   user 20     False
```

2.2.2 如何查询用户及其属性

1. 活跃用户列表的显示

在 Office 365 管理中心左侧的导航栏中单击"用户"按钮，在弹出的扩展菜单上单击"活跃用户"按钮。此时，除非我们应用了某个视图或者在"搜索用户"文本框中键入了某些筛选条件，否则，出现在右侧窗格内的活跃用户列表就包含当前这个租户中所有的用户的用户名、显示姓名、授权等信息，如图 2-11 所示。

图 2-11

要想进一步查看用户的部门、职务等详细信息，可以选择该用户显示姓名左侧的复选框，并在弹出的用户属性对话框中单击"联系人信息"右侧的"编辑"按钮，如图 2-12 所示。

图 2-12

使用 Get-MsolUser 命令可以在不添加任何参数的情况下，查询 Office 365 的用户。命令和输出结果如下：

O365 PS > Get-MsolUser

```
UserPrincipalName                              DisplayName        isLicensed
-----------------                              -----------        ----------
user06@office365ps.partner.onmschina.cn        user 06            False
user01@office365ps.partner.onmschina.cn        user 01            True
user07@office365ps.partner.onmschina.cn        user 07            False
user05@office365ps.partner.onmschina.cn        user 05            False
user10@office365ps.partner.onmschina.cn        user 10            False
user02@office365ps.partner.onmschina.cn        user 02            False
……
```

利用 PowerShell 的管道操作符可以对输出的结果进行灵活的控制，以 Office 365 管理中心用户列表的显示方式来输出查询结果，例如按照显示姓名的升序对用户进行排列，将显示姓名的字段排列在前，并显示许可证的相关信息等。具体的操作步骤如下：

（1）我们先选择要显示的字段，并调整字段的显示顺序。命令和输出结果如下：

O365 PS > Get-MsolUser | Select-Object DisplayName, UserPrincipalName, Licenses

```
DisplayName    UserPrincipalName                              Licenses
-----------    -----------------                              --------
user 06        user06@office365ps.partner.onmschina.cn        {}
user 01        user01@office365ps.partner.onmschina.cn        {office365ps…}
user 07        user07@office365ps.partner.onmschina.cn        {}
user 05        user05@office365ps.partner.onmschina.cn        {}
user 10        user10@office365ps.partner.onmschina.cn        {}
user 02        user02@office365ps.partner.onmschina.cn        {}
```

（2）再调整显示姓名的排列顺序。命令和输出结果如下：

O365 PS > Get-MsolUser | Select-Object DisplayName, UserPrincipalName, Licenses |
>> Sort-Object DisplayName

```
DisplayName    UserPrincipalName                              Licenses
-----------    -----------------                              --------
user 01        user01@office365ps.partner.onmschina.cn        {office365ps…}
user 02        user02@office365ps.partner.onmschina.cn        {}
user 03        user03@office365ps.partner.onmschina.cn        {}
user 04        user04@office365ps.partner.onmschina.cn        {}
user 05        user05@office365ps.partner.onmschina.cn        {}
user 06        user06@office365ps.partner.onmschina.cn        {}
```

PowerShell 不仅可以做到像 Office 365 管理中心的用户列表那样按照某个字段进行排序，而且还能在指定某个排序字段后，再指定另外一个，甚至第三、第四个排序字段，并使不同的字段显示不相同的升降序结果。命令如下：

```
O365 PS > Get-MsolUser | Select-Object DisplayName, UserPrincipalName, Licenses |
>> Sort-Object @ {Expression="DisplayName"; Ascending=$true}, `
>> @ {Expression="Licenses"; Descending=$true}
```

在这条命令中，我们首先按照 DisplayName 的升序，然后按 Licenses 的降序对结果进行排序。而在 Office 365 管理中心的用户列表中，则仅可以分别对显示姓名或用户名进行排序，并且不能按照许可证的状态进行排序，灵活性明显比不上 PowerShell 命令。

2. 用户属性的自定义显示

（1）在用户属性对话框中，要想查看用户属性的具体内容，例如查看用户的职务、部门、办公室等信息，需要单击"联系人信息"项的"编辑"按钮，并在随后弹出的"编辑联系人信息"页面中进行查看，如图 2-13 所示。

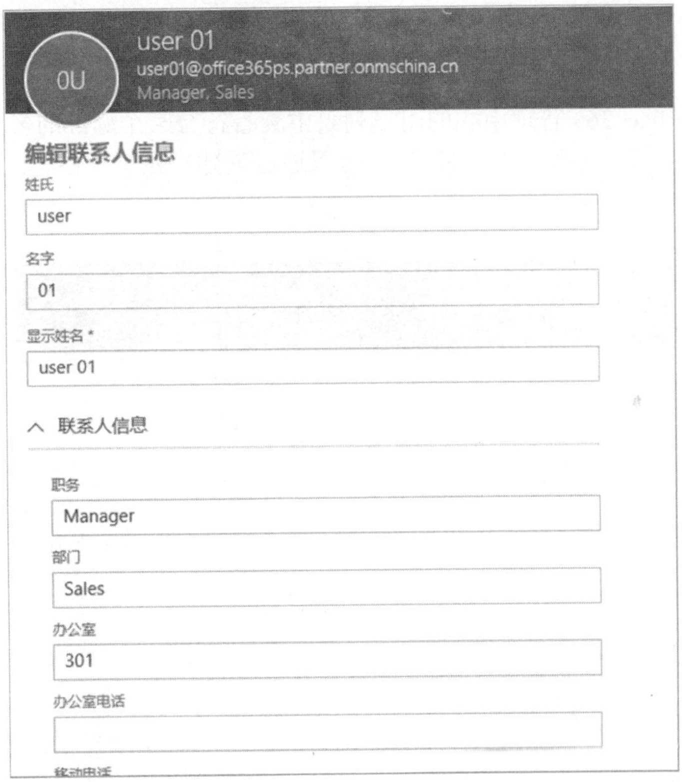

图 2-13

相应地，查询特定用户的属性可以使用以下 PowerShell 命令。命令和输出结果如下：

```
O365 PS > Get-MsolUser -UserPrincipalName "user01@office365ps.partner.onmschina.cn"

UserPrincipalName                              DisplayName    isLicensed
-----------------                              -----------    ----------
user01@office365ps.partner.onmschina.cn        user 01        True
```

如果要查询包含用户 user 01 全部属性的列表，可以使用以下命令。命令和输出结果如下：

O365 PS > Get-MsolUser `
>> -UserPrincipalName "user01@office365ps.partner.onmschina.cn" | Format-List

```
ExtensionData                      : System.Runtime.Serialization.ExtensionDataObject
AlternateEmailAddresses            : {}
AlternateMobilePhones              : {}
AlternativeSecurityIds             : {}
BlockCredential                    : False
City                               :
CloudExchangeRecipientDisplayType  : 1073741824
Country                            :
Department                         : Sales
DirSyncProvisioningErrors          : {}
……
```

（2）要想在 Office 365 管理中心的用户列表中查看符合某个标准的多个用户的信息，例如查看同属于 Sales 部门的用户的头衔信息，可以通过创建自定义视图方式来实现，如图 2-14 所示。

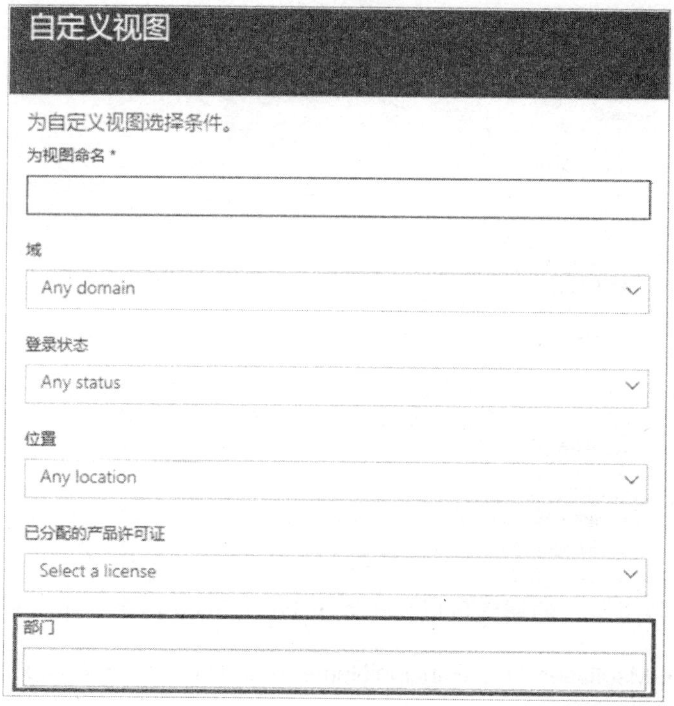

图 2-14

通过自定义视图查找符合某个标准的多个用户，生成的用户列表，内容及内容的排序只能显示三个字段，而且只能以显示姓名、用户名、许可证状态的固定顺序进行排序，如图 2-11

所示。我们关心的其他信息例如用户头衔信息却无法被显示。

为了解决这个问题,我们可以使用以下 PowerShell 命令。这一命令执行的结果只显示我们最关心的显示姓名、部门和头衔这三个字段的内容。命令和输出结果如下:

O365 PS > Get-MsolUser | Where-Object {$_.Department -eq "Sales"} |
>> Select DisplayName, Department,Title | Sort-Object DisplayName

```
DisplayName  Department  Title
-----------  ----------  -----
user 01      Sales       Manager
user 02      Sales
user 03      Sales
```

3. 多重用户筛选标准的使用

在活跃用户列表上,要想找到在显示姓名或用户名里面含有某些关键字的用户,可以在"搜索用户"的文本框中键入这些关键字,然后单击该文本框中放大镜形状的搜索按钮开始查找,如图 2-15 所示。

图 2-15

要想在符合上述条件的用户中进一步查找那些隶属于销售部门的用户,就需要在 Office 365 管理中心的活跃用户列表中额外创建一个自定义的视图。

PowerShell 命令在这方面却要灵活许多,操作步骤如下:

(1)我们先来查询所有用户的情况。命令和输出结果如下:

O365 PS > Get-MsolUser | Select DisplayName, Department | Sort-Object DisplayName

```
DisplayName  Department
-----------  ----------
user 01      Sales
user 02      Sales
user 03      Sales
user 04      Finance
user 05      Finance
user 06      Finance
user 07      HR
user 08      HR
user 09      HR
user 10      Marketing
```

（2）接着在执行命令时新增加两个参数 Department 和 SearchString。命令和输出结果如下：

O365 PS > Get-MsolUser -Department "Sales" -SearchString "user 0" |
>> Select DisplayName, Department | Sort-Object DisplayName

```
DisplayName  Department
-----------  ----------
user 01      Sales
user 02      Sales
user 03      Sales
```

参数 Department 限定了出现在结果集中的用户只能来自销售部门，而参数 SearchString 规定用户的显示姓名中的编号只能以 0 开始，user 10 这个用户将不会出现在结果集中。通过使用 Get-MsolUser 命令并搭配相应的参数，我们可以完成对特定条件的过滤；而使用管道操作符又可以将上述结果集做进一步筛选，或者对显示内容和显示顺序进行调整。

4. 用户查询命令的其他参数

在 Office 365 管理中心的活跃用户列表中，通过单击"视图"下拉列表框，并选择其中的特定视图，我们可以查看在活跃用户中符合特定标准的部分用户的列表，如图 2-16 所示。

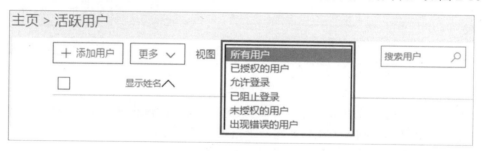

图 2-16

使用 Get-MsolUser 命令并搭配 EnabledFilter 参数，可以查询得到与"所有用户"、"允许登录"和"已阻止登录"用户视图相同的结果集，该参数包括以下属性值：All, EnabledOnly 和 DisabledOnly。命令和输出结果如下：

O365 PS > Get-MsolUser -EnabledFilter DisabledOnly |
>> Select-Object DisplayName, BlockCredential | Sort-Object DisplayName

```
DisplayName  BlockCredential
-----------  ---------------
user 02      True
```

虽然使用 UnlicensedUsersOnly 参数可以查询得到"未授权的用户"的结果集，使用 HasErrorsOnly 参数可得到"出现错误的用户"的结果集，但是要想得到与"已授权的用户"视图相同的结果集，仅凭 UnlicensedUsersOnly 这个参数是无法实现的，因为这是个开关参数，支持 true 和 false 这两个值。使用属性值 true，可以得到所有未授权用户的列表；而使用属性值 false，并不能把已授权的用户显示出来。命令和输出结果如下：

```
O365 PS > Get-MsolUser -UnlicensedUsersOnly:$false |
>> Select-Object DisplayName, isLicensed | Sort-Object DisplayName

DisplayName    IsLicensed
-----------    ----------
user 01        True
user 02        False
user 03        False
user 04        False
user 05        False
user 06        False
user 07        False
user 08        False
user 09        False
user 10        False
```

为了解决这个问题，我们可以使用排除法来查询已经得到授权的用户。命令和输出结果如下：

```
O365 PS > Get-MsolUser | Where-Object {
>> $_.DisplayName -notin (Get-MsolUser -UnlicensedUsersOnly).DisplayName} |
>> Select-Object Displayname,Licenses | Sort-Object Displayname

DisplayName    Licenses
-----------    --------
user 01        {office365ps:O365_BUSINESS_PREMIUM}
```

另外，我们还可以换一种思路来查询已经得到授权的用户。首先，使用管道操作符对 Get-MsolUser 命令的输出结果用 Where-Object 中的条件进行筛选。筛选条件中，"$_.Licenses -ne $null" 检查每条记录的许可证信息是否为空，如果为空将不会包含在最后的输出结果中；然后对结果集进行排序。命令和输出结果如下：

```
O365 PS > Get-MsolUser | Where-Object {$_.Licenses -ne $null} |
>> Select-Object Displayname,Licenses | Sort-Object Displayname

DisplayName    Licenses
-----------    --------
user 01        {office365ps:O365_BUSINESS_PREMIUM}
```

2.2.3 如何修改用户的属性

用户创建完成后，在图形界面中可以修改该用户的属性，步骤如下。

（1）在 Office 365 管理中心左侧的导航栏中单击"用户"按钮，在弹出的扩展菜单上单击"活跃用户"按钮。

（2）在右侧窗格内的活跃用户列表中选择需要修改属性的用户，请参考图 2-9；在页面右侧弹出的包含该用户当前属性的对话框中修改产品许可证、组成员身份等属性的内容，请参考图 2-10。

1. 用户属性修改命令

根据不同的修改需求，我们可以使用下面的命令，并配合使用不同的参数来修改相应的用户属性。

（1）使用 Set-MsolUser 命令修改用户的传真、电话、手机号码等属性。例如 user 01 更换了手机号码，可以使用下面的方式来修改属性。命令如下：

```
O365 PS > Set-MsolUser `
>> -UserPrincipalName "user01@office365ps.partner.onmschina.cn" `
>> -MobilePhone "13812345678"
```

如果命令的执行过程没有报错，属性的修改就完成了。另外，我们还可以使用 Get-MsolUser 命令检查修改后的属性值是否正确。命令和输出结果如下：

```
O365 PS > Get-MsolUser `
>> -UserPrincipalName "user01@office365ps.partner.onmschina.cn" |
>> Select-Object UserPrincipalName,MobilePhone
```

```
UserPrincipalName                              MobilePhone
-----------------                              -----------
user01@office365ps.partner.onmschina.cn        13812345678
```

（2）修改用户名的操作不能使用 Set-MsolUser 命令，而应该使用 Set-MsolUserPrincipalName 命令。我们先创建一个用户，用户标识是 usertest10@office365ps.partner.onmschina.cn。命令和输出结果如下：

```
O365 PS > New-MsolUser `
>> -UserPrincipalName "usertest10@office365ps.partner. onmschina.cn" `
>> -DisplayName "user 10"
```

```
Password   UserPrincipalName                                  DisplayName  isLice
--------   -----------------                                  -----------  ------
E3vYd234   usertest10@office365ps.partner.onmschina.cn        user 10      False
```

然后，我们将这个用户的用户标识从 usertest10@office365ps.partner.onmschina.cn 更改为 user10@office365ps.partner.onmschina.cn。命令如下：

```
O365 PS > Set-MsolUserPrincipalName `
>> -UserPrincipalName "usertest10@office365ps.partner.onmschina.cn"
>> -NewUserPrincipalName "user10@office365ps.partner.onmschina.cn"
```

修改完成以后，用户就可以使用新标识 user10@office365ps.partner.onmschina.cn 登录 Office 365 了，之前设置好的密码及密码策略等均不会受到任何影响。

（3）要为用户修改或重置密码，或者要求用户在下次登录时自行更改其密码，则需要使用 Set-MsolUserPassword 命令，而非 Set-MsolUser 命令。命令和输出结果如下：

O365 PS > Set-MsolUserPassword `
\>> -UserPrincipalName "user05@office365ps.partner.onmschina.cn" `
\>> -NewPassword "Kt3pv669"

```
Kt3pv669
```

O365 PS > Set-MsolUserPassword `
\>> -UserPrincipalName "user06@office365ps.partner. onmschina.cn"

```
Yuz98120
```

上述两条命令分别对应图形管理界面上用户属性对话框中"重置密码"按钮所进行的"创建密码"和"自动生成密码"这两种重置密码的操作。第一条命令有参数 NewPassword，系统将把我们键入的字符串作为用户 user 05 的新密码，并且作为命令的输出结果返回给我们；第二条命令没有 NewPassword 参数，系统将为用户 user 06 设置一个随机的新密码。

上述两条命令成功执行后，用户在首次使用新密码登录时，系统会要求用户进行密码重置，如图 2-17 所示。

图 2-17

如果无须用户在下一次登录的时候重置密码，则只需要添加 ForceChangePassword 参数，并将该参数的值设置为$false 即可。命令如下：

O365 PS > Set-MsolUserPassword `
>> -UserPrincipalName "user07@office365ps.partner.onmschina.cn" `
>> -ForceChangePassword $false -NewPassword "v8905Tq1"

（4）当怀疑一个用户的用户凭据泄露时，为了防止非法用户的登录行为，可以暂时禁止该账户登录 Office 365。命令如下：

O365 PS > Set-MsolUser `
>> -UserPrincipalName "user02@office365ps.partner.onmschina.cn" `
>> -BlockCredential $True

上述命令成功执行后，即使以用户 user 02 正确的用户凭据登录 Office 365，仍然会出现不可登录的提示，如图 2-18 所示。

图 2-18

2. 特定用户属性的修改

在 Office 365 中，为用户设置默认语言或者禁用密码复杂度的策略等，并不只是在新建用户时才能进行设置，也可以通过 PowerShell 命令在用户创建完成后，再进行修改。

（1）用户 user 03 和 user 05 被调去美国总部工作，他们的默认语言需要设置成与总部一致的语言英语。而默认语言的设置是无法由管理员在用户属性对话框中统一进行修改的。使用 PowerShell 命令可以免去用户各自修改该属性的麻烦。命令如下：

O365 PS > $users = @("user03@office365ps.partner.onmschina.cn", `
>> "user08@office365ps. partner.onmschina.cn")
O365 PS > foreach ($user in $users) {
>> Set-MsolUser -UserPrincipalName $user -PreferredLanguage "EN-US" }

（2）在新建用户的时候，管理员可以利用 PowerShell 命令为其设定简单的密码，但是，如果用户的创建已经完成，要将其密码重置为简单密码，是无法在用户属性对话框中进行修改的。这时就需要使用 PowerShell 命令执行以下两个步骤。

① 利用 Set-MsolUser 命令修改用户密码属性为永不过期，并为用户修改密码复杂度的策略。命令如下：

O365 PS > Set-MsolUser `
>> -UserPrincipalName "user04@office365ps.partner.onmschina.cn" `
>> -PasswordNeverExpires $true -StrongPasswordRequired $false

② 使用 Set-MsolUserPassword 命令为该用户重置一个简单密码。命令和输出结果如下：

O365 PS > Set-MsolUserPassword `
>> -UserPrincipalName "user04@office365ps.partner.onmschina.cn" `
>> -ForceChangePassword $false -NewPassword "officeps"

```
officeps
```

2.2.4　如何批量操作用户

如果一个组织的管理员需要维护的用户数量在 20 人以下，使用 PowerShell 命令的操作效率与在 Office 365 管理中心中进行操作看似没有太大的区别，甚至感觉 PowerShell 的操作还不如 Office 365 管理中心图形化的管理界面直观，但是，随着企业规模的扩张，维护规模的扩大，如果管理员需要维护的用户数量已经过百，则使用 PowerShell 命令进行用户维护的优势就展现出来了。

1. 用户的批量添加

我们可以通过以下步骤在 Office 365 管理中心中批量添加用户。

（1）在 Office 365 管理中心左侧的导航栏中单击"用户"按钮，在弹出的扩展菜单上单击"活跃用户"按钮，参见图 2-11。

（2）单击"更多"按钮，并在弹出的扩展菜单中单击"+导入多个用户"按钮，如图 2-19 所示。

PowerShell for Office 365 应用实战

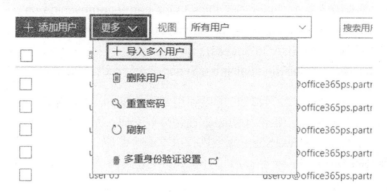

图 2-19

（3）在"导入多个用户"对话框的"创建并上传文件"标签中，单击"下载仅具有标头的 CSV 文件"或者"下载具有标头和示例用户信息的 CSV 文件"按钮，下载模板文件。填写完成并保存后，单击"浏览"按钮，提交刚才保存过的 CSV 文件，然后单击"下一步"按钮，如图 2-20 所示。

图 2-20

> **注意**
>
> 表格里面的"用户名"列，需要采用以下格式：xxx@office365ps.partner.onmschina.cn。这里的域名需要填写本租户中已经有的域名，而不是随便选择一个，否则就会出现报错信息："用户名中使用的域名无效"。

（4）在"导入多个用户"对话框的"设置用户选项"标签中，为用户分配 Office 365 的许可证，如图 2-21 所示。

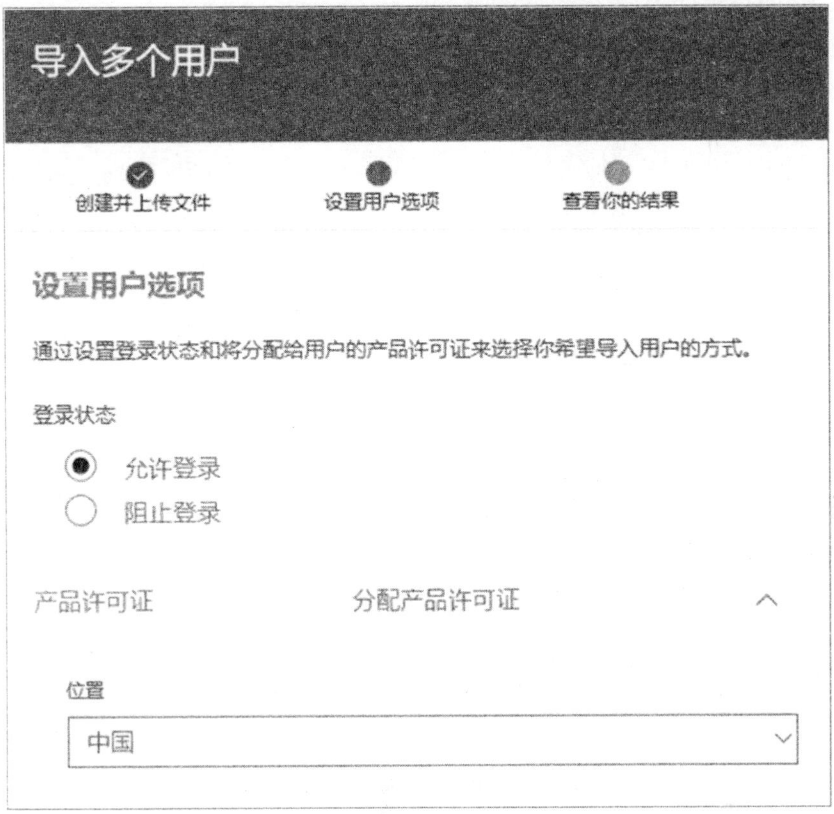

图 2-21

选择许可证的操作完成后，单击"下一步"按钮。在"导入多个用户"对话框的"查看你的结果"标签中确认后，即可完成最终的添加。

批量添加用户的模板文件仅包含以下这些列：用户名、名字、姓氏、显示名称、职务、部门、办公室号码、办公室电话、移动电话、传真号码、地址、城市、省/自治区/直辖市、邮政编码、国家或地区，其中没有包含为用户设置默认语言的属性列。即我们在活跃用户列表中批量添加用户后，需要在用户登录后，由用户自行手动修改默认语言。

而通过使用以下 PowerShell 命令，不仅同样可以批量添加用户，还可以解决上述无法为用户设置默认语言的问题，方法如下。

我们需要先准备一个即将添加用户的列表，不过格式可以由我们自行定义。为了演示方便，并且突出重点，我们只添加包括："用户名""姓氏""名字""显示姓名""部门""使用语言"在内的几个字段。其中，显示名称使用 PowerShell 命令进行拼装，下面是这个列表的部分内容：

user11@office365ps.partner.onmschina.cn,user,11,Sale,zh-CN
user12@office365ps.partner.onmschina.cn,user,12,Sale,en-US
user13@office365ps.partner.onmschina.cn,user,13,Sale,zh-CN
user14@office365ps.partner.onmschina.cn,user,14,HR,en-US

user15@office365ps.partner.onmschina.cn,user,15,HR,zh-CN
user16@office365ps.partner.onmschina.cn,user,16,HR,en-US
user17@office365ps.partner.onmschina.cn,user,17,Finance,zh-CN
user18@office365ps.partner.onmschina.cn,user,18,Finance,en-US
user19@office365ps.partner.onmschina.cn,user,19,Finance,zh-CN
user20@office365ps.partner.onmschina.cn,user,20,Finance,en-US

可以使用 Nodepad++等文本编辑工具，并借助其中的列编辑功能可以提高文件编辑效率。文件准备完成后，我们就可以添加这些用户了。命令和输出结果如下：

```
O365 PS > Get-Content .\userlist.txt | % {
>> $temp = $_.Split(",")
>> $PrincipleN = $temp[0]
>> $LastN = $temp[1]
>> $FirstN = $temp[2]
>> $DisplayN = $LastN+" "+$FirstN
>> $Dep = $temp[3]
>> $PLanguage = $temp[4]
>> New-MsolUser -UserPrincipalName $PrincipleN -FirstName $FirstN `
>> -LastName $LastN -DisplayName $DisplayN -Department $Dep `
>> -PreferredLanguage $PLanguage}

Password  UserPrincipalName                              DisplayName  isLicensed
--------  -----------------                              -----------  ----------
Yz130qaa  user11@office365ps.partner.onmschina.cn        user 11      False
2089qApa  user12@office365ps.partner.onmschina.cn        user 12      False
Eya9p33i  user13@office365ps.partner.onmschina.cn        user 13      False
83e8e6z9  user14@office365ps.partner.onmschina.cn        user 14      False
uiIYop10  user15@office365ps.partner.onmschina.cn        user 15      False
38YTRioi  user16@office365ps.partner.onmschina.cn        user 16      False
Pp3Yp384  user17@office365ps.partner.onmschina.cn        user 17      False
zzUi38ei  user18@office365ps.partner.onmschina.cn        user 18      False
Pqqy82y5  user19@office365ps.partner.onmschina.cn        user 19      False
QeoiqY5I  user20@office365ps.partner.onmschina.cn        user 20      False
```

这里我们使用了 PowerShell 的循环。除了使用循环，我们还使用字符串的 Split()方法，并将以逗号分隔的子字符串分别赋值给了不同的变量；然后，再将这些变量作为参数传递给 New-MsolUser 命令。命令的输出结果显示这些用户已经被创建了。

2. 用户属性的批量修改

作为系统管理员，同时修改多个用户的属性也是家常便饭。例如，我们要把 Finance 部门中所有用户的部门属性统一修改为 Fin，在图形界面上可以采用以下操作步骤。以下。首先，在活跃用户列表上，利用系统已经定义的或自定义的视图将符合标准的用户筛选出来，并逐一选择这些用户显示姓名左侧的复选框。然后，在页面右侧弹出的"批量操作"菜单中单击"编辑联系人信息"按钮，如图 2-22 所示。并在接下来弹出的"编辑联系人信息"对话框中进行部门属性的修改操作。

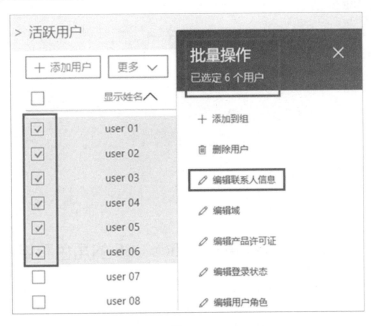

图 2-22

利用 PowerShell 的管道操作符可以将前面命令的输出结果传递给后面的命令做进一步处理，从而高效地实现用户属性的批量修改。另外，我们还可以通过添加筛选条件，更精确地控制要修改的对象。例如，我们可以仅修改用户 user 01 至 user 06 的部门属性，而 Finance 部门中的其他用户不会受到影响。命令如下：

```
O365 PS > Get-MsolUser -SearchString "user" -Department "Finance" |
>> Where-Object {($_.DisplayName -ge "user 01") -and ($_.DisplayName -le "user 06")} |
>> Set-MsolUser -Department "Fin"
```

使用上面这条命令，首先查找部门为 Finance 并且用户名中包含 user 字符串的用户；然后进一步筛选结果集，并限定只包含 user 01 到 user 06。此处注意字符串比较符号 ge、le 及逻辑操作符 and 的使用方法；最后，将符合条件的用户的部门名称从 Finance 修改为 Fin。

接下来，我们使用以下命令验证修改结果。命令和输出结果如下：

```
O365 PS > Get-MsolUser -SearchString "user" | Select-Object DisplayName,Department

DisplayName   Department
-----------   ----------
user 01       Sales
user 02       Sales
user 03       Sales
user 04       Fin
user 05       Fin
user 06       Fin
```

```
user 07    HR
user 08    HR
user 09    HR
user 10    Marketing
user 11    Sale
user 12    Sale
user 13    Sale
user 14    HR
user 15    HR
user 16    HR
user 17    Finance
user 18    Finance
user 19    Finance
user 20    Finance
```

2.3 如何使用 PowerShell 管理 Office 365 的角色和组

除了用户管理，角色和组的管理也是系统管理员的必备技能。因为，随着用户数量的增加，管理的工作量也会增大，所以各种管理职责需要由不同的管理人员承担，这就促使我们把各种管理权限分配出去。用户数量的增加还带来另外一个问题，就是用户管理复杂程度的增加。通过按照不同部门、地域创建不同的组，把用户放到这些组里进行管理可以显著提高管理的效率。

2.3.1 如何管理 Office 365 的角色

1. 角色的查看

随着微软不断添加新的应用到 Office 365 中，新的管理员角色也会随之增加。例如 Teams Service Administrator 目前只能使用 PowerShell 命令进行查询，还未出现在 Office 365 管理中心的角色列表中。在用户属性对话框的"角色"项下，单击"编辑"按钮，并在随后弹出的"编辑用户角色"对话框中，目前只列出了 10 种管理员角色，分为"全局管理员"和"自定义的管理员"两大类，后者又进一步分为 9 种更为具体的角色。而 Office 365 的全部角色列表可以使用下面的 PowerShell 命令获得。命令和输出结果如下：

O365 PS > Get-MsolRole | Select-Object Name

```
Name
----
Device Users
Device Administrators
Device Join
Workplace Device Join
Directory Synchronization Accounts
Device Managers
```

```
User Account Administrator
Helpdesk Administrator
Service Support Administrator
Billing Administrator
Partner Tier2 Support
Exchange Service Administrator
SharePoint Service Administrator
Lync Service Administrator
Compliance Administrator
Company Administrator
Partner Tier1 Support
Directory Readers
Directory Writers
Application Administrator
Application Developer
Security Reader
Security Administrator
Privileged Role Administrator
Intune Service Administrator
Cloud Application Administrator
Customer LockBox Access Approver
CRM Service Administrator
Power BI Service Administrator
Guest Inviter
Authentication Administrator
Privileged Authentication Administrator
Teams Communications Administrator
Teams Communications Support Engineer
Teams Communications Support Specialist
Teams Service Administrator
```

我们将命令结果中的这 36 个角色进行分类，并对重点的角色进行详细的介绍。

（1）全局管理员是租户中具有最高特权的管理员角色，这个角色的成员可以查看、配置或修改这个租户中所有应用的属性，也可以将其他管理员角色分配给其他用户。按照最佳管理实践，应该将全局管理员角色分配给企业中的多个用户。因为，如果租户中仅有的一个全局管理员因为各种情况无法登录时，由于其拥有最高的管理权限，所以只有全局管理员才能进行管理工作便无法正常进行。要想为其他用户分配全局管理员角色，操作者本身首先也要是一个全局管理员。

在 Office 365 管理中心的 Office 365 角色列表的"编辑用户角色"对话框中，我们可以看到的角色名称是"全局管理员"或 Global Administrator，而使用 Get-MsolRole 命令列出的全部角色列表中却找不到这个角色。实际上，全局管理员在使用 PowerShell 命令为其他用户分配该角色的时候，该角色对应的角色名称是 Company Administrator。为其他用户分配全局管理员角色的命令如下：

O365 PS > Add-MsolRoleMember -RoleName "Company Administrator" `

>> -RoleMemberEmailAddress "user10@office365ps.partner.onmschina.cn"

上述命令成功执行后，用户 user 10 会被分配全局管理员的角色。我们在使用 PowerShell 命令为用户分配角色时，要留意角色和角色名称之间的差异。密码管理员这个角色在 PowerShell 中的角色名称是 Helpdesk Administrator。

（2）有一类角色通过角色的名称就能了解它们的作用，它们是管理 Office 365 各项应用的管理员角色，拥有在各自应用中的管理员权限，可对相关应用中的设置进行修改，例如，Exchange Service Administrator、SharePoint Service Administrator、Lync Service Administrator、Power BI Service Administrator、Teams Service Administrator、CRM Service Administrator 等。

（3）有一类角色虽然不是某项应用的管理员角色，但是它们承担了某些公共的管理职责，例如对用户进行管理，对账单进行管理等。表 2-1 中列举了其中几个重要的角色。

表 2-1

角 色	职 权
Service Support Administrator	创建微软服务支持工单，查看服务状态和消息中心的内容
Billing Administrator	采购、管理订阅，监控服务健康状态，管理支持工单
User Account Administrator	添加删除用户账号、组，重置用户密码，监控服务健康状态，但是无法删除其他管理员角色中的成员，或为他们重置密码
Helpdesk Administrator	为非特权用户和其他同角色中的成员重置密码，管理微软服务支持工单，查看服务状态
Privileged Role Administrator	进行角色的指派管理，包括特权角色成员的管理
Authentication Administrator	查看非特权用户认证方法的信息，以及为他们设定或重置认证方法
Privileged Authentication Administrator	查看用户认证方法信息，以及为他们设定或重置认证方法，包括非特权以及特权的用户
Compliance Administrator	管理组织的合规策略，在管理主页，Exchange Online 管理中心主页都有权限
Security Administrator	管理组织的安全策略，在管理主页，Exchange Online 管理中心主页都有权限

（4）角色 Customer LockBox Access Approver 很少被提起，为了能了解这个角色的作用，我们首先需要知道什么是 Lockbox Requests。当用户使用 Office 365 遇到问题时，微软的技术支持工程师为了进行排错，有可能会接触到用户的数据，而 LockBox Requests 就是让用户可以对微软工程师访问其数据的行为进行控制，但是，并不是每次 Office 365 问题排错都会用到 LockBox Access，如果微软工程师没有机会接触到用户的数据，就没有 LockBox Access 的问题。角色 Customer LockBox Access Approver 就是为批准 LockBox Requests 而设立的。另外，LockBox 只出现在 Office 365 E5 的产品订阅中。

（5）角色 Device Managers 已经不再使用了，而 Partner Tier1 Support 和 Partner Tier2 Support 这两个角色不是通用角色，我们也不要轻易使用。

2. 角色的分配

（1）在 Office 365 管理中心中为用户分配管理员角色的步骤如下。

① 在 Office 365 管理中心左侧的导航栏中单击"用户"按钮，在弹出的扩展菜单上单击"活跃用户"按钮。

② 在活跃用户列表中选择要分配管理员角色的用户。

③ 在页面右侧弹出的用户属性对话框中的"角色"项中单击"编辑"按钮。

④ 在弹出的"编辑用户角色"对话框中，通过选择"全局管理员"或选择"自定义的管理员"各个角色左侧的复选框，将管理员角色分配给该用户，如图 2-23 所示。

图 2-23

要想同时给多个用户分配相同的管理员角色，我们就需要在"活跃用户"的列表中，选择那些即将被赋予管理员角色的用户显示姓名左侧的复选框；然后，在页面右侧弹出的"批量操作"菜单中单击"编辑用户角色"按钮，如图 2-24 所示。

图 2-24

在使用 PowerShell 命令为多个用户分配相同的管理员角色时，我们要注意，如果我们将多个用户名以逗号分隔字符串的形式赋给 RoleMemberEmailAddress 参数，会收到以下报错信息。命令和输出结果如下：

O365 PS > Add-MsolRoleMember -RoleName "Helpdesk Administrator" `
\>\> -RoleMemberEmailAddress "user18@office365ps.partner.onmschina.cn, `
\>\> user19@office365ps.partner.onmschina.cn, `

```
            >> user20@office365ps.partner.onmschina.cn"
```

```
Add-MsolRoleMember : Access Denied. You do not have permissions to call this
cmdlet.
At line:1 char:1
+ Add-MsolRoleMember -RoleName "Helpdesk Administrator" -RoleMemberEmai ...
+ ~~~~~~~~~~~~~~~~~~~~~~~~~~~~~~~~~~~~~~~~~~~~~~~~~~~
    + CategoryInfo          : OperationStopped: (:) [Add-MsolRoleMember],
MicrosoftOnlineException
    + FullyQualifiedErrorId                                                    :
Microsoft.Online.Administration.Automation.UserNotFoundException,Microsoft.On
line.Administration.Automation.AddRoleMember
```

上述问题并不是因为权限不足导致的，因为我们创建到 Office 365 租户的连接时，使用的是全局管理员的用户凭据。

我们可以使用 PowerShell 循环结构来解决上述问题。命令如下：

O365 PS > $MemberEmail = @("user18@office365ps.partner.onmschina.cn", `
>> "user19@office365ps.partner.onmschina.cn", `
>> "user20@office365ps.partner.onmschina.cn")

O365 PS > $MemberEmail | % {Add-MsolRoleMember `
>> -RoleName "Helpdesk Administrator" -RoleMemberEmailAddress $_}

首先把用户的邮件地址放到一个字符串数组中，接着再利用循环结构为他们逐一分配密码管理员的角色。

（2）在 Office 365 管理中心的活跃用户列表中，我们无法很直观地进行角色的批量撤销操作。当我们选择被分配了相同或不同管理角色的多个用户显示姓名左侧的复选框，并在页面右侧弹出的"批量操作"菜单中单击"编辑用户角色"按钮，弹出"编辑角色"对话框后，在该对话框中，会显示角色依然是"用户（没有管理员访问权限）"，如图 2-25 所示。

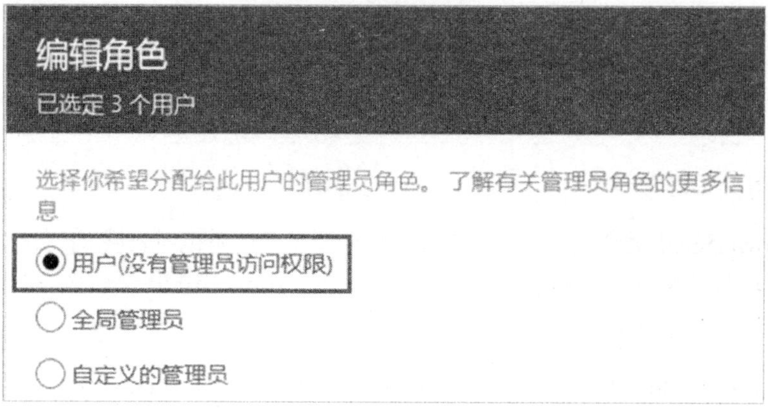

图 2-25

这时，虽然这些用户在对话框中显示没有被赋予管理员的角色，但是如果我们单击"保存"按钮，分配给这些用户的所有角色都将被撤销。为了避免误操作，我们可以一次只选中一个用户，在页面右侧弹出的用户属性对话框中的"角色"项下，单击"编辑"按钮，然后在弹出的"编辑用户角色"对话框中，选择"用户（没有管理员访问权限）"选项，来撤销该用户的管理员角色。

使用 Remove-MsolRoleMember 命令来进行用户角色的批量撤销操作，相比而言可控性更好。命令如下：

```
O365 PS > $MemberEmail = @("user18@office365ps.partner.onmschina.cn", `
>> "user19@ office365ps.partner.onmschina.cn")

O365 PS > $MemberEmail | % {Remove-MsolRoleMember `
>> -RoleName "Helpdesk Administrator"-RoleMemberType User `
>> -RoleMemberEmailAddress $_}
```

因为 RoleMemberEmailAddress 参数不接受逗号分隔形式的字符串，我们还是用 PowerShell 循环来进行批量撤销的操作，最后我们可以用 Get-MsolRoleMember 命令确认删除的结果。

（3）在 Office 365 管理中心中很难查看被分配了特定管理员角色的用户，例如，有哪些用户被分配了密码管理员的角色。使用 PowerShell 命令却可以轻松搞定。这里有一点需要注意，Get-MsolRoleMember 命令只接受 RoleObjectId 参数，而不是 RoleName，所以需要将 RoleName 转化成 RoleObjectId。命令和输出结果如下：

```
O365 PS > Get-MsolRoleMember `
>> -RoleObjectId (Get-MsolRole -RoleName "Helpdesk Administrator").ObjectId |
>> Select-Object EmailAddress, DisplayName

EmailAddress                                DisplayName
------------                                -----------
user18@office365ps.partner.onmschina.cn     user 18
user19@office365ps.partner.onmschina.cn     user 19
user20@office365ps.partner.onmschina.cn     user 20
```

2.3.2 如何管理 Office 365 的安全组

1. 安全组的管理命令

打开 Office 365 管理中心，在左侧的导航栏中单击"组"按钮，在弹出的扩展菜单中选择"组"，如图 2-26 所示。

图 2-26

（1）在组列表页面单击"+添加组"按钮，然后在页面右侧弹出的"添加组"对话框中，通过下拉列表框选择"安全"组类型，并键入组名称，最后单击"添加"按钮添加组，如图 2-27 所示。

图 2-27

在添加组的对话框中创建安全组时，我们无法指定组的所有者（创建 Office 365 类型的组的时候，可以同时选择组的所有者）。只有在该组创建后，在组列表中选择这个安全组的组名左侧的复选框，并在页面右侧弹出的组属性对话框中单击"所有者"项下的"编辑"按钮，才能添加组的所有者，如图 2-28 所示。

图 2-28

相比而言，使用 PowerShell 命令添加安全组的同时就可以方便地为这个安全组指定所有者。注意：用 ManagedBy 参数指定组的所有者时，如果使用了用户的显示名称或用户的电子邮件地址，均会报如下的错误。命令和输出结果如下：

O365 PS > New-MsolGroup -DisplayName "secHR" `
>> -ManagedBy "admin@office365ps.partner.onmschina.cn"

```
New-MsolGroup : Invalid value for parameter. Parameter name: ManagedBy.
At line:1 char:1
+ New-MsolGroup -DisplayName "secHR" -ManagedBy "admin@office365ps.part ...
+ ~~~~~~~~~~~~~~~~~~~~~~~~~~~~~~~~~~~~~~~~~~~~~~~~
    + CategoryInfo          : OperationStopped: (:) [New-MsolGroup],
MicrosoftOnlineException
    + FullyQualifiedErrorId : Microsoft.Online.Administration.Automation.
PropertyValidationException,Microsoft.Online.Administration.Automation.NewGro
up
```

使用之前介绍过的 Get-MsolUser 命令获取即将成为该组管理员的用户对象标识，接着把这个对象标识作为参数值赋给 ManagedBy 参数，可以解决以上问题。命令的输出结果显示 secHR 安全组已经建立。命令和输出结果如下：

O365 PS > New-MsolGroup -DisplayName "secHR" -ManagedBy (Get-MsolUser `
>> -UserPrincipalName "admin@office365ps.partner.onmschina.cn").ObjcectId

```
ObjectId                             DisplayName       GroupType      Description
--------                             -----------       ---------      -----------
4041f612-b6b3-49c2-b533-7423d49a1f16  secHR            Security
```

注意

如果仅安装了 Microsoft Azure Active Directory Module for Windows PowerShell 模块，使用模块中的 New-MsolGroup 命令，是无法创建"Office 365"、"通讯组列表"和"启用邮

件的安全"这三种类型的组的。至于如何创建这三种类型的组,我们会在后面使用 PowerShell 管理 Exchange Online 的部分进行介绍。

(2)在组列表上,可以通过选择"查看"下拉列表框中的组类型来显示不同类型组的列表,而通过在"搜索"文本框中键入部分或完整的组名来查找特定名称的组,如图 2-29 所示。

图 2-29

若要用 PowerShell 命令的方式来获取租户内的组信息,可以仅使用 Get-MsolGroup 命令,不带任何参数,来列出租户中所有的组。命令和输出结果如下:

O365 PS > Get-MsolGroup

```
ObjectId                              DisplayName       GroupType            Description
--------                              -----------       ---------            -----------
4041f612-b6b3-49c2-b533-7423d49a1f16  secHR             Security
150ef228-b080-41d0-b2a9-18c68ecac9e5  disSales          DistributionList
8827185c-08c3-422c-b1fe-4687ede5f526  sdiSales          MailEnabledSecurity
8ece97a8-860e-4ad2-a932-7e9c49f68c03  sdiHR             MailEnabledSecurity
```

注意:与 New-MsolGroup 命令只能用来创建新的安全组不同,Get-MsolGroup 命令不仅会列出本租户中的安全组,还会一并列出所有其他类型的组,而我们可以用 GroupType 这个参数来控制结果中仅显示所有的安全组。命令和输出结果如下:

O365 PS > Get-MsolGroup -GroupType Security

```
ObjectId                              DisplayName       GroupType            Description
--------                              -----------       ---------            -----------
7499d8cc-c48c-47f6-9314-8052b7a2257f  secSales          Security
4041f612-b6b3-49c2-b533-7423d49a1f16  secHR             Security
```

在 Office 365 管理中心的组列表上的搜索文本框中键入部分组名进行搜索的时候,这个搜索字符串需要从特定组名称的头部开始进行截取,否则,将不会返回正确的搜索结果。类似地,如果使用 Get-MsolGroup 命令搭配 SearchString 参数查找有特定名称的组时,也需要保证这个被赋给 SearchString 参数的字符串是从特定组名称的头部开始截取的。然而,如果我们换一种思路来解决这个问题，即可让该搜索字符串是名称字符串中的任意子集。命令和输出结果如下：

```
O365 PS > Get-MsolGroup -GroupType Security |
>> Where-Object { $_.DisplayName -match "HR"}

ObjectId                              DisplayName        GroupType    Description
--------                              -----------        ---------    -----------
4041f612-b6b3-49c2-b533-7423d49a1f16  secHR              Security
9b2867e8-1854-4be4-ae22-0cf2ff5a577b  testHR             Security
```

（3）我们可以使用 Remove-MsolGroup 命令来移除不再需要的安全组，唯一需要注意的是，被移除的组是用组的 ObjectId 来表示的，而非组的名称。命令和输出结果如下：

```
O365 PS > Remove-MsolGroup `
>> -ObjectId (Get-MsolGroup -SearchString "secHR").ObjectId

Confirm
Continue with this operation?
[Y] Yes  [N] No  [S] Suspend  [?] Help (default is "Y"): Y
```

2. 安全组的组成员管理

（1）在 Office 365 管理中心中向特定的安全组添加成员的方法如下：在组列表页面选择该安全组的组名左侧的复选框，接着在页面右侧弹出的组属性对话框中单击成员项下的"编辑"按钮，并在随后弹出的组成员管理对话框中单击"+添加成员"按钮，如图2-30所示。

图 2-30

在选择组的成员时，可以在"搜索以添加成员"文本框中键入用户标识或用户的显示姓名，来快速定位需要添加到这个安全组的成员，如图 2-31 所示。

图 2-31

以上方法只能按照用户名进行搜索。要想把销售部的所有同事添加到 secSales 组，我们事先需要按照每个用户所属的组对用户进行分类，然后才能在成员搜索文本框中通过逐个搜索用户名的这种方式来进行添加。如果销售部的同事比较多，将会带来不小的工作量。

使用 Add-MsolGroupMember 命令配合相应的条件筛选，可以很方便地实现上述需求。但是我们要注意，这条命令的 GroupObjectId 参数只接受安全组的对象标识，而不是安全组的名称。命令如下：

```
O365 PS > $grpID = (Get-MsolGroup -SearchString "secSales").ObjectId
O365 PS > Get-MsolUser | ForEach-Object { if($_.Department -match "Sales") {
>> $userID = $_.ObjectID
>> Add-MsolGroupMember -GroupObjectId $grpID `
>> -GroupMemberType user -GroupMemberObjectId $userID}
>> }
```

我们先把安全组的对象标识存储在$grpID 这个变量中，然后通过 Get-MsolUser 命令查询本租户中的所有用户，并查看每个用户的部门名称，如果该用户属于 Sales 部门，就将其添加到 secSales 组中。在上述命令片段中我们使用了管道操作符，用来对本租户中的所有用户应用组的筛选条件，并将符合条件的用户添加到 secSales 组。

（2）通过在 Office 365 管理中心的组列表中选择特定组的组名左侧的复选框，然后在页面右侧弹出的组属性对话框中查看成员项下包括的用户，可以确认该特定安全组中包括的成员，如图 2-32 所示。

第 2 章 使用 PowerShell 管理 Office 365 的租户

图 2-32

若要用 PowerShell 命令的方式来查看安全组的成员，使用 Get-MsolGroupMember 命令即可。类似地，我们只能使用安全组的对象标识符，而不能使用安全组的组名。命令和输出结果如下：

O365 PS > Get-MsolGroupMember `
>> -GroupObjectId (Get-MsolGroup -SearchString "secSales").ObjectId

```
GroupMemberType EmailAddress                                DisplayName
--------------- ------------                                -----------
User            user01@office365ps.partner.onmschina.cn     user 01
User            user02@office365ps.partner.onmschina.cn     user 02
User            user03@office365ps.partner.onmschina.cn     user 03
User            user11@office365ps.partner.onmschina.cn     user 11
User            user12@office365ps.partner.onmschina.cn     user 12
User            user13@office365ps.partner.onmschina.cn     user 13
```

如果要查看特定用户隶属于哪些安全组，可以在 Office 365 管理中心的活跃用户列表中选择用户显示姓名左侧的复选框选择该用户，然后在页面右侧弹出的用户属性对话框中查看组成员身份项下列出的组，即可确定该用户是哪些安全组的成员，如图 2-33 所示。

图 2-33

以下 PowerShell 脚本片段，使用了两个循环来实现类似的功能，外层循环列出了租户中的每个安全组，并使用 Get-MsolGroupMember 命令查询每个组里面包含的用户；内层循环通过用户标识，判断在当前组中是否存在目标用户，如果存在该目标用户，就输出这个组名。其中的条件判断使用了字符串的 StartsWith()方法来判断当前用户的邮件地址是否以$user 这个变量开头。命令和输出结果如下：

```
O365 PS > $user = "user03"
O365 PS > Get-MsolGroup -GroupType Security | ForEach-Object {
>> $grpNAME = $_.DisplayName
>> $grpID = $_.ObjectId
>> Get-MsolGroupMember -GroupObjectId $grpID | ForEach-Object {
>> $userEMAIL = $_.EmailAddress
>> if ($userEMAIL.StartsWith($user)) {
>> Write-Host $grpNAME}
>> }
>> }

secSales
```

（3）在 Office 365 管理中心中，我们可以通过编辑用户属性对话框中的组成员身份或者编辑组属性对话框中的成员列表，来实现删除特定安全组中的成员用户的目的。相应地，我们可以使用 Remove-MsolGroupMember 命令来删除特定安全组中的成员，要注意的是，使用这条命令时，GroupObjectId、GroupMemberType、GroupMemberObjectId 这三个参数是必选的。命令如下：

```
O365 PS > Remove-MsolGroupMember `
>> -GroupObjectId (Get-MsolGroup -SearchString "secSales").ObjectId `
>> -GroupMemberType User -GroupMemberObjectId (Get-MsolUser `
>> -UserPrincipalName "user03@office365ps.partner.onmschina.cn").ObjectId
```

2.4 如何使用 PowerShell 管理 Office 365 的订阅和许可证

除了用户、角色、组，这些 Office 365 管理中一定会涉及的对象，系统管理员还必须关注 Office 365 的订阅和许可证。

订阅和许可证，并不是同一个事物的两种称谓，而是互相关联的两种事物。当我们注册一个新租户的时候，如果不购买订阅，企业里的用户是无法使用 Word、Excel、PowerPoint、邮件、即时通信等 Office 365 的组件和应用的。企业一般根据组织所需使用的 Office 365 的组件和应用，以及企业中的用户数量来确定所需购买的订阅类型和许可证数量，当企业购买了不同版本的 Office 365 订阅后，系统管理员会在创建用户账户的时候或者用户账户创建后，将许可证分配给用户，并可以购买更多的许可证来满足新入职人员使用产品和服务的需求。如果有人员离职，系统管理员可以删除该用户的许可证，并分配给其他用户使用。总之，订阅是企业按时间和需求对 Office 365 的产品和应用使用权的购买，而许可证是为了方便计算企业中可以使用的 Office 365 产品和服务的数量。

2.4.1 如何查看订阅及相关信息

1. 订阅和许可证的查看

（1）在 Office 365 管理中心中单击左侧导航栏中的"账单[①]"按钮，然后在弹出的扩展菜单中单击"订阅"按钮，即可在右侧详细内容窗格内查看租户的订阅信息了，如图 2-34 所示。

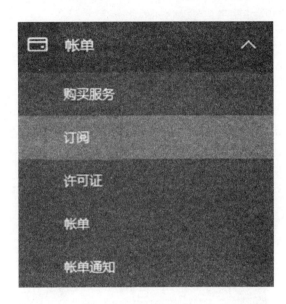

图 2-34

相应地，使用 PowerShell 的 Get-MsolSubscription 命令也可以查询订阅种类、许可证到期日、订阅目前状态、许可证总数等信息。命令和输出结果如下：

O365 PS > Get-MsolSubscription |
>> Select-Object SkuPartNumber,NextLifecycleDate,Status,TotalLicenses

```
SkuPartNumber         NextLifecycleDate      Status  TotalLicenses
-------------         -----------------      ------  -------------
O365_BUSINESS_PREMIUM 5/9/2019 11:59:59 PM   Enabled            25
```

而 OcpSubscriptionId、SkuId 等信息，只能通过使用 PowerShell 命令才能查询到。命令和输出结果如下：

O365 PS > Get-MsolSubscription | Format-List

---

① 图中"帐单"应为"账单"，本书正文使用"账单"。

```
    ExtensionData          : System.Runtime.Serialization.ExtensionDataObject
    DateCreated            : 4/3/2018 12:00:00 AM
    IsTrial                : True
    NextLifecycleDate      : 5/9/2019 11:59:59 PM
    ObjectId               : xxx-xxx
    OcpSubscriptionId      : xxx-xxx
    OwnerObjectId          :
    OwnerType              :
    ServiceStatus          : {Microsoft.Online.Administration.ServiceStatus,
Microsoft.Online.Administration.ServiceStatus,
                             Microsoft.Online.Administration.ServiceStatus,
Microsoft.Online.Administration.ServiceStatus...}
    SkuId                  : xxx-xxx
    SkuPartNumber          : O365_BUSINESS_PREMIUM
    Status                 : Enabled
    TotalLicenses          : 25
```

（2）在 Office 365 管理中心中单击左侧导航栏中的"账单"按钮，然后在弹出的扩展菜单中单击"许可证"按钮，即可查看租户许可证的使用情况，如图 2-35 所示。

图 2-35

相应地，使用不带参数的 Get-MsolAccountSku 命令，同样可以查询本租户许可证的使用情况，或者配合管道操作符和 Format-List 参数获取许可证使用的完整情况。命令和输出结果如下：

O365 PS > Get-MsolAccountSku

```
AccountSkuId                            ActiveUnits  WarningUnits  ConsumedUnits
------------                            -----------  ------------  -------------
office365ps:O365_BUSINESS_PREMIUM 25                 0             5
```

2. 账单通知用户的查看

在 Office 365 管理中心中单击左侧导航栏中的"账单"按钮，接着在弹出的扩展菜单中单击"账单通知"按钮，即可在右侧的窗格内具体查看有哪些用户可以接收账单通知，如图 2-36 所示。

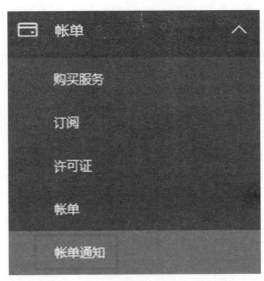

图 2-36

租户中可以购买许可证和接收账单通知的管理员角色只有两种——全局管理员和账务管理员。表 2-2 记录了不同管理员角色可以进行的有关许可证的相关操作。

表 2-2

管理员角色	分配许可证	删除许可证	购买更多许可证	删除账户
全局管理员	是	是	是	是
账务管理员	否	否	是	否
用户管理管理员	是	是	否	是
服务管理员	否	否	否	否
密码管理员	否	否	否	不支持

要想用 PowerShell 命令查询可以接收账单通知的用户，我们可以先将能接收账单通知的两类管理员的角色名称放到字符串数组中，接着利用 PowerShell 循环结构依次查询被分配了两类管理员角色的用户账户，并将他们的邮件地址显示出来。命令和输出结果如下：

O365 PS > $billROLE = @("Company Administrator","Billing Administrator")
O365 PS > $billROLE | ForEach-Object {
>> Get-MsolRoleMember -RoleObjectId (Get-MsolRole -RoleName $_).ObjectId |
>> Select-Object EmailAddress}

```
EmailAddress
------------
admin@office365ps.partner.onmschina.cn
user10@office365ps.partner.onmschina.cn
user01@office365ps.partner.onmschina.cn
```

2.4.2 如何为用户分配许可证

只有为用户分配许可证后，用户才能使用 Office 365 的组件和应用。既可以在新建用户的同时直接分配许可证，也可以在用户创建以后再分配许可证。

1. 新建用户许可证的分配

在 Office 365 管理中心的活跃用户列表中单击"+添加用户"按钮后，在页面右侧弹出的"新建用户"对话框中，"产品许可证"项目是必须设置的。即使我们暂时不准备给用户分配许可证，也需要手动开启"创建用户但不分配产品许可证"开关，如图 2-37 所示。

图 2-37

在使用 PowerShell 命令 New-MsolUser 创建新用户时，如果不使用 LicenseAssignment 参数，默认是不会分配任何许可证给该新用户的。在管理中心的活跃用户列表中，用户的授权状态也会显示为未授权。如果配合使用 LicenseAssignment 参数，则可以在创建用户的同时给用户分配许可证。注意，该参数的值一般是"租户:订阅类型"这种格式。另外，我们可以使用 Get-MsolAccountSku 命令获得该参数的属性值。命令和输出结果如下：

O365 PS > Get-MsolAccountSku

```
AccountSkuId                          ActiveUnits  WarningUnits  ConsumedUnits
------------                          -----------  ------------  -------------
office365ps:O365_BUSINESS_PREMIUM  0           25            23
```

O365 PS > New-MsolUser -UserPrincipalName "user21@office365ps.partner.onmschina.cn" `

```
>> -DisplayName "user 21" `
>> -LicenseAssignment "office365ps:O365_BUSINESS_PREMIUM"
```

```
New-MsolUser : You must provide a required property: Parameter name:
UsageLocation
At line:1 char:1
+ New-MsolUser -UserPrincipalName "user21@office365ps.partner.onmschina ...
+ ~~~~~~~~~~~~~~~~~~~~~~~~~~~~~~~~~~~~~~~~~~~~~~~~~~~~~~~~~~~~~~~~
    + CategoryInfo          : OperationStopped: (:) [New-MsolUser],
MicrosoftOnlineException
    + FullyQualifiedErrorId : Microsoft.Online.Administration.Automation.
RequiredPropertyNotSetException,Microsoft.Online.Administration.Automation.Ne
wUser
```

此处的报错是因为上述命令缺少 UsageLocation 参数。在通过 Office 365 管理中心的"新用户"对话框新建用户时，用户位置属性已经被设置为"中国"；而在使用 LicenseAssignment 参数的时候，我们需要配合使用 UsageLocation 这个参数，参数值是国家代码，中国为 CN。命令和输出结果如下：

```
O365 PS > New-MsolUser -UserPrincipalName "user21@office365ps.partner.onmschina.cn" `
>> -DisplayName "user 21" -UsageLocation "CN" `
>> -LicenseAssignment "office365ps:O365_BUSINESS_PREMIUM"
```

```
Password  UserPrincipalName                          DisplayName isLicensed
--------  -----------------                          ----------- ----------
zEE38902  user21@office365ps.partner.onmschina.cn    user 21     True
```

2. 现有用户许可证的添加

如果在新建用户的时候没有为其分配许可证，之后为他们添加许可证的时候，我们可以在 Office 365 管理中心的活跃用户列表中，选择该用户显示姓名左侧的复选框，接着在页面右侧弹出的对话框中单击"产品许可证"项下的"编辑"按钮，如图 2-38 所示。

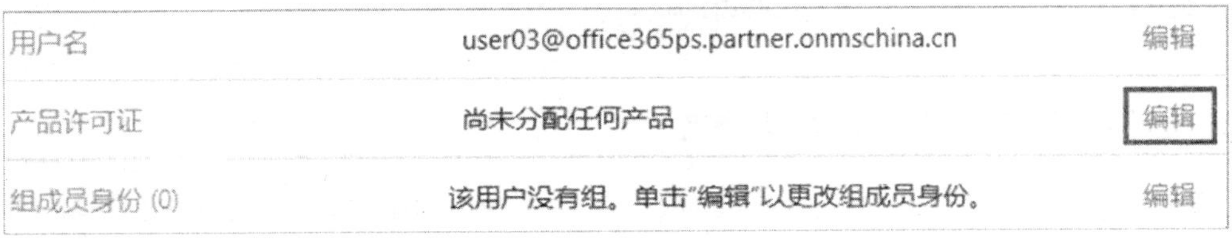

用户名	user03@office365ps.partner.onmschina.cn	编辑
产品许可证	尚未分配任何产品	编辑
组成员身份 (0)	该用户没有组。单击"编辑"以更改组成员身份。	编辑

图 2-38

注意

在创建用户时没有分配许可证，用户创建完成后为这些用户添加许可证时，不能使用 Set-MsolUser 命令，而需要使用 Set- MsolUserLicense 命令。

在分配许可证前要先确认这个用户的 UsageLocation 属性不是空值，如果该属性是空值，也会出现必选属性缺失的报错。命令和输出结果如下：

```
O365 PS > Set-MsolUserLicense `
>> -UserPrincipalName "user02@office365ps.partner. onmschina.cn" `
>> -AddLicenses "office365ps:O365_BUSINESS_PREMIUM"
```

```
Set-MsolUserLicense : You must provide a required property: Parameter name:
UsageLocation
   At line:1 char:1
   + Set-MsolUserLicense -UserPrincipalName "user02@office365ps.partner.on ...
   + ~~~~~~~~~~~~~~~~~~~~~~~~~~~~~~~~~~~~~~~~~~
       + CategoryInfo          : OperationStopped: (:) [Set-MsolUserLicense],
MicrosoftOnlineException
       + FullyQualifiedErrorId : Microsoft.Online.Administration.Automation.
RequiredPropertyNotSetException,Microsoft.Online.Administration.Automation.Se
tUserLicense
```

我们可以提前使用 PowerShell 命令判断用户的 UsageLocation 属性是否已经进行过设置。如果没有被设置，就在设置完用户的 UsageLocation 属性后，再为该用户添加许可证。命令和输出结果如下：

```
O365 PS > $userEMAIL = "user02@office365ps.partner.onmschina.cn"
O365 PS > if ((Get-MsolUser -UserPrincipalName $userEMAIL).UsageLocation -eq $null) {
>> Set-MsolUser -UserPrincipalName $userEMAIL -UsageLocation "CN"}
O365 PS > Set-MsolUserLicense `
>> -UserPrincipalName "user02@office365ps.partner. onmschina.cn" `
>> -AddLicenses "office365ps:O365_BUSINESS_PREMIUM"
O365 PS > Get-MsolUser -UserPrincipalName $userEMAIL | Select-Object Licenses
```

```
Licenses
--------
{office365ps:O365_BUSINESS_PREMIUM}
```

2.4.3 如何对许可证进行定制化修改

Office 365 的订阅包括整套标准的 Office 365 产品和应用。按照订阅版本的不同，提供了一定程度的差异化服务。我们虽然无法对订阅中的产品和应用进行更改，但是可以在向用户分配许可证的时候，单击 Office 365 管理中心中的用户属性对话框"产品许可证"项下的"编辑"按钮，然后在"产品许可证"对话框中，使用拨动选择开关按钮，来禁用或开启订阅中具体的服务或应用，如图 2-39 所示。

图 2-39

在使用 PowerShell 命令对许可证进行修改时，通过创建许可证选项对象，在给用户分配或添加许可证时，同样可以调整许可证中具体的服务或应用的开闭状态，从而达到定制化修改的目的。另外，我们可以借助循环，高效地为一组用户执行批量操作。

1. 许可证中各应用状态的查看

查看许可证中各应用状态的命令和输出结果如下：

O365 PS > $user = "user02@office365ps.partner.onmschina.cn"
O365 PS > (Get-MsolUser -UserPrincipalName $user).Licenses.ServiceStatus

```
ServicePlan             ProvisioningStatus
-----------             ------------------
Microsoft Bookings      Success
SHAREPOINTWAC           Success
SHAREPOINTSTANDARD      Success
OFFICE_BUSINESS         Success
MCOSTANDARD             Success
EXCHANGE_S_STANDARD     Success
```

从查询命令的输出结果我们可以看到，用户 user 02 目前可以使用许可证中所有的服务和应用。

2. 许可证中应用的禁用和恢复

（1）禁用。

新建一个许可证选项，并在该选项中禁用 Skype for Business Online。命令如下：

O365 PS > $opt = New-MsolLicenseOptions `
>> -AccountSkuId office365ps:O365_ BUSINESS_PREMIUM `
>> -DisabledPlans MCOSTANDARD

然后，使用 Set-MsolUserLicense 命令为用户 user 02 应用上述许可证选项。命令如下：

O365 PS > Set-MsolUserLicense -UserPrincipalName $user -LicenseOptions $opt

成功执行上述命令后，再次查看该用户许可证中各服务和应用的状态，可以看到 Skype for Business Online 已经被成功禁用了。命令和输出结果如下：

O365 PS > (Get-MsolUser -UserPrincipalName $user).Licenses.ServiceStatus

```
ServicePlan              ProvisioningStatus
-----------              ------------------
Microsoft Bookings       Success
SHAREPOINTWAC            Success
SHAREPOINTSTANDARD       Success
OFFICE_BUSINESS          Success
MCOSTANDARD              Disabled
EXCHANGE_S_STANDARD      Success
```

（2）恢复。

首先使用以下命令来重置许可证选项：

O365 PS > $opt = New-MsolLicenseOptions `
>> -AccountSkuId "office365ps:O365_BUSINESS_ PREMIUM"

然后为用户 user 02 应用上述重置过的许可证选项。命令如下：

O365 PS > Set-MsolUserLicense -UserPrincipalName $user -LicenseOptions $opt

最后，再次验证修改的结果。命令和输出结果如下：

O365 PS > (Get-MsolUser -UserPrincipalName $user).Licenses.ServiceStatus

```
ServicePlan              ProvisioningStatus
-----------              ------------------
Microsoft Bookings       Success
SHAREPOINTWAC            Success
SHAREPOINTSTANDARD       Success
OFFICE_BUSINESS          Success
MCOSTANDARD              Success
EXCHANGE_S_STANDARD      Success
```

注意

许可证中服务和应用状态的变更需要一些时间，如果在状态变更的过程中执行 Get-MsolUser 命令，ProvisioningStatus 会被显示成 PendingInput。

第 3 章 使用 PowerShell 管理 Office 365 的 Exchange Online

在现代企业里，使用电子邮件进行沟通已经是企业运作不可或缺的部分了，但是对于系统管理员而言，管理一个自有的企业邮件系统来保证各项业务的正常运转，确实不是一件简单的事情，因为我们要面对邮箱备份是否完成，垃圾邮件如何处理等一系列的问题；而使用 Office 365 的 Exchange Online，系统管理员就无须花费太多的精力在关注服务器硬件的健康状态、进行备份升级等基础的运维工作上，可以把更多的时间放在用户需求的管理上，例如创建邮箱、创建通讯录、管理手机访问等。另外，通过使用 PowerShell 来进行 Exchange Online 的管理，会让系统管理员的日常管理工作更加高效便捷。

3.1 如何使用 PowerShell 连接 Exchange Online

当企业规模变得越来越大时，创建不同角色的管理员（例如 Skype for Business Online 的管理员、SharePoint Online 的管理员）分担不同的管理任务，是最佳的权限管理实践。本章就用被分配了 Exchange Online 管理员角色的用户 exo Admin 来进行管理操作。

1. 使用 PowerShell 连接 Exchange Online 的具体步骤

在 Office 365 管理中心中单击左侧导航栏中的"管理中心"按钮，然后在弹出的扩展菜单中，单击 Exchange 按钮进入 Exchange 管理中心，如图 3-1 所示。

由于用户 exo Admin 仅仅被分配了 Exchange Online 管理员的角色，之前使用全局管理员用户凭据登录时，"管理中心"扩展菜单中有的 Skype for Business、SharePoint、Azure Active Directory 等按钮都不再显示了。另外，Exchange 管理中心可以使用以下网址直接登录：

图 3-1

https://partner.outlook.cn/ecp/

使用 PowerShell 管理 Exchange Online 并不需要安装额外的管理模块，创建一个与 Exchange Online 的连接就可以了。首先，打开一个 PowerShell 控制台窗口，然后键入并执行以下命令，命令和输出结果如下：

```
EXO PS > $cred= Get-Credential
EXO PS > $uri = "https://partner.outlook.cn/PowerShell"
EXO PS > $Session = New-PSSession `
>> -ConfigurationName Microsoft.Exchange -ConnectionUri $uri `
>> -Credential $cred -Authentication Basic -AllowRedirection

WARNING: Your connection has been redirected to the following URI:
"https://partner.outlook.cn/PowerShell-LiveID?PSVersion=5.1.17134.765 "

EXO PS > Import-PSSession $Session -DisableNameChecking

ModuleType Version    Name                    ExportedCommands
---------- -------    ----                    ----------------
Script     1.0                                tmp_34uvhhhy.t3f
{Add-AvailabilityAddressSpace, Add-DistributionGroupMember, Add-Mail...
```

第一条命令用来获取并保存用户凭据。执行的时候会弹出一个对话框，提示用户键入用户名和密码，我们就用 Exchange 管理员 exo Admin 的邮件地址作为用户名。

第二条命令将指定 Exchange Online 的统一资源标识符，即连接端点的域名地址。域名是以".cn"结尾的，代表了连接中国区的 Exchange Online；国际版地址则是以".com"结尾的；德国版是以".de"结尾的。

第三条命令创建了一个会话，并为这个会话指定以下参数：参数 ConnectionUri 使用中国版 Exchange Online 的统一资源标识符，参数 Credential 引用了我们存储的用户凭据，参数 Authentication 用于定义用户凭据的验证方式，而参数 AllowRedirection 允许我们把连接重定向到备用的统一资源标识符。

注意

与本书第 2 章中使用 Connect-MsolService 连接 Office 365 的租户不同，连接 Exchange Online 时，需要使用创建并导入会话的方式。创建和导入会话是有先后顺序的，只有创建了一个会话后，才能导入它。

最后一条命令，即导入之前创建的远程会话，并使前面命令中创建的隐式远程模块中包含的命令和方法在这个会话中可用。在命令执行的时候，PowerShell 控制台窗口上方会出现具体模块的加载进度提示，如图 3-2 所示。

```
Creating implicit remoting module ...
    Getting command information from remote session ... 211 commands received
    [oooooooooooooooooooooooooooooooooo
    00:00:07 remaining.
```

图 3-2

如果不添加 DisableNameChecking 参数，会出现下面的警告提示（但是可以忽略）：
WARNING: The names of some imported commands from the module

' tmp_34uvhhhy.t3f ' include unapproved verbs that might make them less discoverable.

如果执行了 2 次 Import-PSSession 命令，会报错，具体的报错信息如下：

Import-PSSession : No command proxies have been created, because all of the requested remote commands would shadow existing local commands.

上述报错只是提示我们，不会再次导入目前会话中已有的同名命令了。不过，我们还是可以使用 AllowClobber 参数强制重新加载这些同名的命令。

由于创建的隐式远程模块的名称每次都是不同的，除了执行 Import-PSSession 命令可以查看这个远程模块的名称外，还可以执行 Get-Module 命令来查询模块的名称。命令和输出结果如下：

EXO PS > Get-Module

```
ModuleType Version     Name                                ExportedCommands
---------- -------     ----                                ----------------
Manifest   3.1.0.0     Microsoft.PowerShell.Management     {Add-Computer,
Add-Content, Checkpoint-Computer, Clear-Content...}
Manifest   3.0.0.0     Microsoft.PowerShell.Security
{ConvertFrom-SecureString, ConvertTo-SecureString, Get-Acl, Get-Auth...
Manifest   3.1.0.0     Microsoft.PowerShell.Utility        {Add-Member,
Add-Type, Clear-Variable, Compare-Object...}
Script     1.2                                             PSReadline
{Get-PSReadlineKeyHandler, Get-PSReadlineOption, Remove-PSReadlineKe...
Script     1.0                                             tmp_34uvhhhy.t3f
{Add-AvailabilityAddressSpace, Add-DistributionGroupMember, Add-Mail...
```

当我们有了模块的名称，就可以进一步查询这个远程模块的其他信息了，例如该隐式远程模块中包含的命令数量等信息。命令和输出结果如下：

EXO PS > $exoModule = Get-Module -Name "tmp_34uvhhhy.t3f"

EXO PS > $exoModule.ExportedCommands.Count

729

2. 使用 PowerShell 连接 Exchange Online 的常见问题

（1）如果我们使用网络搜索引擎查询到的是国际版或者德国版本的 URI，而非中国版本 Exchange Online 的 URI 时，就会报以下错误：

[FailureCategory=Cafe-HttpProxyException] Failed to resolve tenant from SMTP address。

所以，一定要使用中国区 Exchange Online 的统一资源标识符来连接，就是带".cn"结尾的那个 URI。

（2）使用 PowerShell 连接 Exchange Online 时，使用的是创建并导入会话这种连接方式。这种方式最大的好处是，每当我们创建新的连接时，对模块的更新就会被及时加载，使我们总是能用到最新版本的管理模块，但是这种建立永久连接的方式，同一时间只允许数量有限的并行连接。如果我们不及时中断不再使用的连接，该连接还会继续保持一段时间的激活状态；如果连接的数量达到上限，管理员就无法创建新的连接了。当我们执行 Import-PSSession 命令后，在结果信息里就会包含这个会话中创建的远程模块的名称，例如 tmp_3aihdi13.pd0、tmp_ynan205e.ukc 等。如果同一个 Exchange Online 管理员同时打开多个 PowerShell 的控制台窗口，每个控制台窗口里面都运行连接 Exchange Online 的 PowerShell 命令，当该管理员在第四个控制台窗口中尝试执行新建会话的 New-PSSession 命令时，就会收到下面的警告信息：

[FailureCategory=AuthZ-AuthorizationException] Fail to create a runspace because you have exceeded the maximum number of connections allowed : 3 for the policy party : MaxConcurrency. Please close existing runspace and try again.

在一个已经开启了 Exchange Online 远程会话的 PowerShell 控制台窗口中，我们可以利用以下命令来获取连接中的远程会话。命令和输出结果如下：

EXO PS > Get-PSSession | Where-Object {$_.ComputerName -eq 'partner.outlook.cn'}

```
  Id Name            ComputerName        ComputerType     State
ConfigurationName    Availability
  -- ----            ------------        ------------     -----
-----------------    ------------
   1 Session1        partner.outl...     RemoteMachine    Opened
Microsoft.Exchange   Available
```

注意

这里获取到的会话只是当前控制台窗口中的会话，其他控制台窗口中的会话是无法显示在这个列表中的。如果我们没有使用 Remove-PSSession 主动退出这个连接中的会话，而是直接关闭这个控制台窗口，这个会话还会保持一段时间的激活状态，而这个未关闭的"孤儿会话"依然会占用并行会话的连接数量，并导致新会话创建失败。

如果我们已经创建了一个 Exchange Online 的会话，但是有很长一段时间没有进行任何操作，并且这个会话的控制台窗口中没有关闭，那么在下一次我们键入 PowerShell 命令并按下回车键后，会出现下面的这条提示：

Creating a new session for implicit remoting of "Get-Mailbox" command...

而且，会再次弹出获取用户凭据的窗口，并等待我们再次键入密码。如果身份验证正常通过，我们刚键入的那条 PowerShell 命令也会继续执行。

因此，管理 Exchange Online 时要及时关闭不再需要的会话，来释放占用的资源。使用下面的命令就可以关闭本控制台窗口内所有的 Exchange Online 会话：

EXO PS > Get-PSSession | Where-Object {$_.ComputerName -eq 'partner.outlook.cn'} |
>> Remove-PSSession

（3）如果键入了错误的用户名和密码，我们是可以从下面的报错信息中发现一些端倪的：
[FailureCategory=LiveID-InvalidCreds] Access Denied
如果遇到上述的错误，请重新尝试键入用户凭据。另外，也可以使用下面的命令显示凭

据的 UserName 和 Password 来进行确认。命令和输出结果如下：

EXO PS > $cred.UserName

```
exoadmin@office365ps.partner.onmschina.cn
```

EXO PS > $cred.GetNetworkCredential().Password

```
Office365PS
```

（4）与 Skype for Business Online 不同，用户即使不是 Exchange Online 管理员，也可以使用 PowerShell 连接 Exchange Online。如果经过验证，键入的用户名和密码是正确的，却依然无法正常创建会话，则可以请租户的全局管理员或者其他 Exchange Online 管理员在连接 Exchange Online 的远程模块后使用以下命令查询该用户的下列属性。命令和输出结果如下：

EXO PS > Get-User -Identity "exoadmin@office365ps.partner.onmschina.cn" |
>> Select-Object RemotePowerShellEnabled

```
RemotePowerShellEnabled
-----------------------
                   True
```

如果这个属性的值是 False，可以使用以下命令修改该属性的值，并在修改完成后使用 Get-User 命令再次进行验证。然而，此处由于用户 exo Admin 该属性的值已经是 True，所以有命令成功完成但是属性没有被修改的警告消息。命令和输出结果如下：

EXO PS > Set-User -Identity "exoadmin@office365ps.partner.onmschina.cn" `
>> -RemotePowerShellEnabled $True

```
WARNING: The command completed successfully but no settings of 'exoAdmin' have
been modified.
```

3. 使用 PowerShell 解决每次都需要输入用户凭据的烦恼

每当我们需要使用用户凭据的时候都会使用 Get-Credential 命令，并在弹出的对话框中键入用户名和密码。如果我们想要免去每次都输入密码的麻烦，可以考虑先键入一个日后可以作为密码的字符串，并把这个作为密码的字符串转换成加密字符串，存储成一个文本文件。下次需要使用密码的时候，直接读取这个文本文件，并将加密的字符串和预先设定的用户名转换成用户凭据，即可用来登录了。

（1）我们使用下面的命令生成并保存加密的字符串。命令和输出结果如下：

EXO PS > $plainPW = Read-Host | ConvertTo-SecureString -AsPlainText -force

```
Office365PS
```

EXO PS > $plainPW | ConvertFrom-SecureString | Out-File .\Password.txt

第一条命令将提示我们键入一个字符串作为密码，例如键入 Office365PS，当按下回车键后，这个字符串会被转换成加密字符串。第二条命令先将这个加密字符串转换成加密的标准字符串，并存储于一个普通的文本文件中，以方便下次读取使用，该文本文件中的内容就如图 3-3 所示。

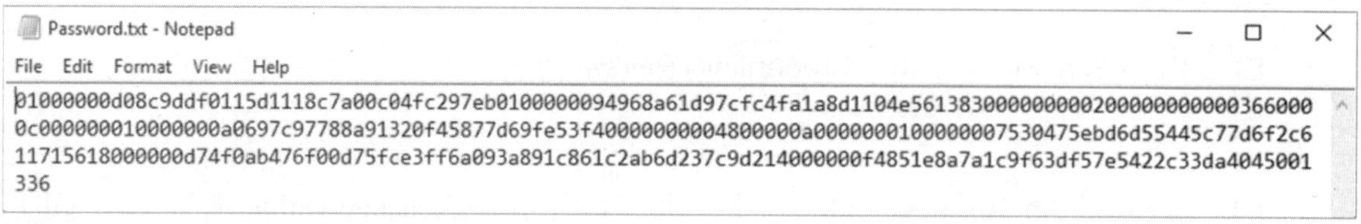

```
01000000d08c9ddf0115d1118c7a00c04fc297eb0100000094968a61d97cfc4fa1a8d1104e561383000000000200000000000366000
0c0000000100000000a0697c97788a91320f45877d69fe53f40000000004800000a000000001000000075304ebd6d55445c77d6f2c6
117156180000000d74f0ab476f00d75fce3ff6a093a891c861c2ab6d237c9d214000000f4851e8a7a1c9f63df57e5422c33da4045001
336
```

图 3-3

（2）我们尝试读取并还原这个加密的标准字符串，并转换成用户凭据对象。命令如下：

EXO PS > $PW = Get-Content .\Password.txt | ConvertTo-SecureString
EXO PS > $cred = New-Object `
>> -TypeName System.Management.Automation.PSCredential `
>> ("exoadmin@office365ps.partner.onmschina.cn", $PW)

在第二条命令中，我们将用户名和转换回加密字符串的密码作为参数，生成了一个用户凭据。为了达到验证目的，下面的命令分别列出了这个凭据包含的用户名和密码，我们可以看到和最初键入的密码字符串是完全一致的。命令和输出结果如下：

EXO PS > $cred.UserName

exoadmin@office365ps.partner.onmschina.cn

EXO PS > $cred.GetNetworkCredential().Password

```
Office365PS
```

3.2 如何使用 PowerShell 管理 Exchange Online 的收件人

作为 Exchange Online 管理员，收件人管理是日常管理工作中的重头戏。当企业中的员工数量增加或者员工信息变更的时候，我们就要花费时间和精力进行收件人的增删改查操作。使用 PowerShell 来进行收件人管理，将帮助我们简化这些日常工作，并提高工作的效率。我们这里说的收件人，并不是狭义的收件人，它不仅包括拥有邮箱的用户，还包括组、资源、联系人等一系列的对象。

3.2.1 如何管理邮箱

1．用户邮箱的创建

（1）企业邮箱已经成为几乎所有员工的标准配置了，我们需要为新同事在 Office 365 中进行一些账户的初始化工作，例如新建用户账户、分配许可证等，接下来就是为这些新同事创建邮箱。然而，Exchange Online 管理员却无法在 Exchange 管理中心进行用户邮箱的创建工作，更不用说创建邮箱的同时为用户创建 Office 365 的账号了。在 Exchange 管理中心中单击左侧导航栏中的"收件人"按钮，然后在右侧详细内容窗格中单击"邮箱"标签按钮，我们会发现没有添加邮箱的按钮，如图 3-4 所示。

图 3-4

即使 Exchange Online 管理员登录 Office 365 管理中心，在活跃用户列表菜单栏中对全局管理员可见的"+添加用户"按钮，对 Exchange Online 管理员也是隐藏的，如图 3-5 所示。

图 3-5

虽然 Exchange Online 管理员无法使用图形界面为用户创建邮箱，但是却可以使用 PowerShell 命令 New-Mailbox 进行用户邮箱的创建工作。命令和输出结果如下：

EXO PS > New-Mailbox -Name "user31" `
>> -MicrosoftOnlineServicesID "user31@office365ps.partner.onmschina.cn"

```
Parameter set cannot be resolved using the specified named parameters.
    + CategoryInfo                    : InvalidArgument: (:) [New-Mailbox],
ParameterBindingException
    + FullyQualifiedErrorId : AmbiguousParameterSet,New-Mailbox
    + PSComputerName        : partner.outlook.cn
```

这里的报错，并不是因为使用 New-Mailbox 命令无法进行用户邮箱的创建，而是使用这条命令需要同时为新用户指定密码。对比本书第 2 章中的 New-MsolUser 命令，命令 New-MsolUser 如果不指定密码，将会生成一个随机的密码。

> **注意**
>
> 这里使用的密码对象必须是加密字符串，需要用 ConvertTo-SecureString -String 'Office365PS' -AsPlainText -Force 这种方式来生成，也可以直接从键盘键入密码字符串 Read-Host -AsSecureString 来生成。我们还可以使用之前介绍过的方法，从文本文件获取转换后的加密字符串，并再次转换为密码对象的方法，来为用户提供密码 Get-Content .\Password.txt | ConvertTo-SecureString。

虽然我们指定了密码，但是还可以通过添加 ResetPasswordOnNextLogon 参数，让该用户在首次登录 Office 365 的时候，自己更改密码。命令和输出结果如下：

```
EXO PS > New-Mailbox -Name "user31" `
>> -MicrosoftOnlineServicesID "user31@ office365ps.partner.onmschina.cn" `
>> -Password (ConvertTo-SecureString -String 'Office365PS' -AsPlainText -Force) `
>> -FirstName "user" -LastName "31" -DisplayName "user 31"

WARNING: After you create a new mailbox, you must go to the Office 365 Admin
Center and assign the mailbox a license, or it will be disabled after the
grace period.

Name                      Alias                         Database                   
ProhibitSendQuota    ExternalDirectoryObjectId
----                      -----                         --------
-----------------    -----------------------
user31                    user31                        CHNPR01DG028-db008         99 GB
(106,300,44...  be09c72f-7b6f-4c34-93fb-67f398b0...
WARNING: Failed to replicate mailbox:'user31' to the
site:'CHNPR01.prod.partner.outlook.cn/Configuration/Sites/SH2PR01'. The
mailbox will be available for logon in approximately 15 minutes.
```

这条命令设定 user31@office365ps.partner.onmschina.cn 为用户的身份标识，在不使用 DisplayName 参数的情况下，会按照 Name 这个参数的值来设定该账户的显示名称。

（2）如果上述命令成功执行，我们在 Office 365 管理中心的活跃用户列表中就能查询到 user 31 这个用户了，即除了用户邮箱被创建，用户账户 user 31 同时也被隐式地创建了。我们等待 15 分钟后，就可以使用 user 31 的用户凭据正常登录 Office 365 了，而且向这个邮件地址发送的邮件也不会被退信，但是，我们要注意执行创建邮箱命令时的警告：如果在缓冲期过后这个账户仍然没有许可证，账户将会被禁用。

在 Office 365 管理中心中，当 Exchange Online 管理员选择活跃用户列表中用户 user 31 左侧的复选框后，页面右侧将弹出用户属性对话框，然后单击"产品许可证"项下的"查看详细信息"按钮，我们会看到 Exchange Online 管理员是没有为用户添加许可证的权限的，如图 3-6 所示。

第 3 章 使用 PowerShell 管理 Office 365 的 Exchange Online

图 3-6

这就需要 Exchange Online 管理员与企业的用户管理管理员或全局管理员协作进行许可证的添加工作。Office 365 在创建邮箱时设置了 grace period 缓冲期，也使不同管理员之间协同工作成为可能。接下来，我们新打开一个 PowerShell 的控制台窗口，并以用户管理管理员的用户凭据登录，使用 Connect-MsolService 连接 Office 365 的租户，为新用户 user 31 分配许可证（此处为用户分配的是"Office 365 商业高级版"订阅的许可证）。另外，我们在给用户分配许可证之前，需要先为用户指定 UsageLocation，并在许可证分配完成后进行检查和确认。命令和输出结果如下：

O365 PS > Set-MsolUser `
>> -UserPrincipalName "user31@office365ps.partner.onmschina.cn" `
>> -UsageLocation "CN"
O365 PS > Set-MsolUserLicense `
>> -UserPrincipalName "user31@office365ps.partner.onmschina.cn" `
>> -AddLicenses "office365ps:O365_BUSINESS_PREMIUM"
O365 PS > Get-MsolUser `
>> -UserPrincipalName "user31@office365ps.partner. onmschina.cn" |
>> Select-Object Licenses

```
Licenses
--------
{office365ps:O365_BUSINESS_PREMIUM}
```

使用以上步骤创建用户邮箱和账户后，用户 user 31 即可正常登录 Office 365，并能打开网页版本的 Outlook，正常地收发邮件了。在给该用户分配许可证前，向该用户的邮件地址发送的测试邮件，也可以在该用户的收件箱中被查看到，说明在新建用户邮箱的缓冲期内，即使没有给用户分配许可证，用户的邮箱也可以正常地接收邮件。

（3）在实际工作中，一次性地为多个用户添加邮箱的情况很常见。通过使用 PowerShell 脚本可以为一个新用户添加邮箱，再结合使用循环语句，即可高效地为多个用户添加邮箱。在本书的第 2 章中，我们介绍了使用逗号分隔的文本文件，批量添加用户账户的方法。接下来，我们来看看使用 csv 文件作为数据源，批量为新用户添加邮箱的步骤。

我们先来准备保存新用户邮箱信息的 csv 文件。具体的步骤是新建一个 Excel 文档，在首行加入列标题，例如 Name、Password 等，接着在第二行开始的后续行中键入新用户的邮箱信息、显示名称等内容。文件中的密码字段键入明文密码即可，我们会在 PowerShell 脚本中对它进行处理；最后，在保存成 ".csv" 格式文件的过程中，系统会提示：一些特性不可用。我们可以忽略该警告，填写完成后的 csv 文件如图 3-7 所示。

	A	B	C	D	E	F
1	Name	ServicesID	Password	FirstName	LastName	DisplayName
2	user33	user33@office365ps.partner.onmschina.cn	Office365	user	33	user 33
3	user34	user34@office365ps.partner.onmschina.cn	Office365	user	34	user 34
4	user35	user35@office365ps.partner.onmschina.cn	Office365	user	35	user 35

图 3-7

接下来，我们就可以批量添加用户的邮箱了。首先，需要使用 Import-Csv 命令导入 CSV 格式的文件。命令和输出结果如下：

EXO PS > $mailBoxes = Import-Csv .\NewMailbox.csv
EXO PS > $mailBoxes | Format-Table

```
Name    ServicesID                                Password  FirstName LastName
----    ----------                                --------  --------- --------
user33  user33@office365ps.partner.onmschina.cn   Office365 user      33
user34  user34@office365ps.partner.onmschina.cn   Office365 user      34
user35  user35@office365ps.partner.onmschina.cn   Office365 user      35
```

然后，使用循环批量地添加用户邮箱。命令和输出结果如下：

EXO PS > foreach ($mb in $mailBoxes) {
>> New-Mailbox -Name $mb.Name -MicrosoftOnlineServicesID $mb.ServicesID `
>> -Password (ConvertTo-SecureString -String "$($mb.Password)" -AsPlainText -Force) `
>> -FirstName $mb.FirstName -LastName $mb.LastName `
>> -DisplayName $mb.DisplayName -ResetPasswordOnNextLogon}

```
WARNING: After you create a new mailbox, you must go to the Office 365 Admin
Center and assign the mailbox a license, or it will be disabled after the
grace period.
```

```
Name                                        Alias                      Database                          
ProhibitSendQuota    ExternalDirectoryObjectId
----                                        -----                      --------
-----------------    -------------------------
user33                                      user33                     CHNPR01DG028-db040                99 GB
(106,300,44...   27724661-83c9-4939-b0c0-48f6c942...
WARNING: Failed to replicate mailbox:'user 33' to the
site:'CHNPR01.prod.partner.outlook.cn/Configuration/Sites/SH2PR01'. The
mailbox will be available for logon in approximately 15 minutes.
WARNING: After you create a new mailbox, you must go to the Office 365 Admin
Center and assign the mailbox a license, or it will be disabled after the grace
period.
user34                                      user34                     CHNPR01DG028-db104                99 GB
(106,300,44...   bc9116f6-86ce-4c84-a368-90f18891...
……
```

这里使用了 foreach 循环，每次将 CSV 格式数据$mailBoxes 中的一行赋给$mb 对象，并在$mb 对象中使用列标题获得每个属性的值，然后使用 New-Mailbox 命令逐个添加用户的邮箱。

> **注意**
>
> 逗号分隔的文本文件没有列标题，而 CSV 格式的文件是包含列标题的。而且，使用 Import-Csv 命令导入的 CSV 格式数据的内容可以直接使用列标题来引用，例如：$mb.Name。

最后，由用户管理管理员或者全局管理员批量设置用户的 UsageLocation，并分配许可证。命令和输出结果如下：

O365 PS > $mailBoxes = Import-Csv .\NewMailbox.csv
O365 PS > foreach ($mb in $mailBoxes) {
>> Set-MsolUser -UserPrincipalName $mb.ServicesID -UsageLocation "CN"
>> Set-MsolUserLicense -UserPrincipalName $mb.ServicesID `
>> -AddLicenses "office365ps:O365_BUSINESS_PREMIUM"}
O365 PS > foreach ($mb in $mailBoxes) {
>> Get-MsolUser -UserPrincipalName $mb.ServicesID |
>> Select-Object DisplayName, Licenses}

```
DisplayName  Licenses
-----------  --------
user 33      {office365ps:O365_BUSINESS_PREMIUM}
user 34      {office365ps:O365_BUSINESS_PREMIUM}
user 35      {office365ps:O365_BUSINESS_PREMIUM}
```

2. 用户邮箱的查看

（1）使用 Get-Mailbox 命令查看当前租户中的邮箱，可以不带任何参数。命令和输出结果如下：

```
EXO PS > Get-Mailbox

Name                                           Alias                       Database             ProhibitSendQuota    ExternalDirectoryObjectId
----                                           -----                       --------             -----------------    -------------------------
DiscoverySearchMailbox...  DiscoverySea...  CHNPR01DG019-db057    50 GB (53,687,091...
exoAdmin                   exoAdmin         CHNPR01DG015-db111    49.5 GB (53,150,2... c7eb42fc-0124-4137-9bc4-a9b1086f...
admin                      admin            CHNPR01DG007-db103    99 GB (106,300,44... 9a4ee2ca-5fa7-4a22-af5a-97717c9f...
sfbadmin                   sfbadmin         CHNPR01DG018-db069    49.5 GB (53,150,2... c7c0bd65-5cc5-44b4-a11e-2f1a3146...
spoadmin                   spoadmin         CHNPR01DG015-db018    49.5 GB (53,150,2... 144503af-5e3c-4491-842b-727a6253...
user01                     user01           CHNPR01DG004-db011    49.5 GB (53,150,2... 122aac1c-5e46-492c-add6-c5cad66e...
……
```

要想按照 Exchange 管理中心邮箱列表的默认样式，仅显示邮箱类型为 User Mailbox（用户）的邮箱，可以通过添加筛选条件来实现。命令和输出结果如下：

```
EXO PS > Get-Mailbox | Where-Object {$_.RecipientTypeDetails -eq "UserMailbox"} |
>> Select-Object DisplayName,RecipientTypeDetails,MicrosoftOnlineServicesID

DisplayName   RecipientTypeDetails   MicrosoftOnlineServicesID
-----------   --------------------   -------------------------
exo Admin     UserMailbox            exoAdmin@office365ps.partner.onmschina.cn
FrankCai      UserMailbox            admin@office365ps.partner.onmschina.cn
sfb Admin     UserMailbox            sfbadmin@office365ps.partner.onmschina.cn
spo Admin     UserMailbox            spoadmin@office365ps.partner.onmschina.cn
user 01       UserMailbox            user01@office365ps.partner.onmschina.cn
user 02       UserMailbox            user02@office365ps.partner.onmschina.cn
user 21       UserMailbox            user21@office365ps.partner.onmschina.cn
user 31       UserMailbox            user31@office365ps.partner.onmschina.cn
user 32       UserMailbox            user32@office365ps.partner.onmschina.cn
user 33       UserMailbox            user33@office365ps.partner.onmschina.cn
user 34       UserMailbox            user34@office365ps.partner.onmschina.cn
user 35       UserMailbox            user35@office365ps.partner.onmschina.cn
```

（2）随着企业中的员工越来越多，用户邮箱的数量也会不断增加，搜索特定邮箱这一功能就变得不可或缺了。在 Exchange 管理中心中单击左侧导航栏中的"收件人"按钮，然后在右侧详细内容窗格中单击"邮箱"标签按钮，如果在用户邮箱列表上方的菜单栏中没有搜索文本框，可以单击放大镜形状的"搜索"按钮使搜索文本框弹出；在搜索文本框中键入需要查找的字符串，然后单击放大镜形状的"搜索邮箱"按钮后，即可在用户邮箱列表中查看结果了,如图 3-8 所示。

图 3-8

但是,当我们要查找 user 31 这个用户,并在搜索文本框中键入要搜索的字符串"31"时,却无法查询到任何的结果。原因就在于,Exchange 管理中心的"搜索邮箱"功能使用的是 Anr 这个模糊查询参数。命令和输出结果如下:

EXO PS > Get-Mailbox -Anr "admin" |
>> Select-Object DisplayName,RecipientTypeDetails, MicrosoftOnlineServicesID

```
DisplayName  RecipientTypeDetails  MicrosoftOnlineServicesID
-----------  --------------------  -------------------------
spo Admin    UserMailbox           spoadmin@office365ps.partner.onmschina.cn
exo Admin    UserMailbox           exoAdmin@office365ps.partner.onmschina.cn
sfb Admin    UserMailbox           sfbadmin@office365ps.partner.onmschina.cn
FrankCai     UserMailbox           admin@office365ps.partner.onmschina.cn
```

参数 Anr 只能在通用名称、显示名称、别名、姓和名等属性中查找以被搜索字符串开头的用户属性值,而且该参数还不支持以通配符开始的字符串。如果我们在用户邮箱列表上方菜单栏的搜索文本框中键入了以"*"开始的搜索字符串,然后单击放大镜形状的"搜索邮箱"按钮,会收到以下提示信息,如图 3-9 所示。

图 3-9

虽然用户邮箱列表上方菜单栏中的"高级搜索"功能,可以在更多的用户属性中进行搜索,但是该功能仍然不支持以通配符开始的搜索字符串,即以"*"开始的搜索字符串;而使用 PowerShell 命令,并配合 Identity 参数,则可以在邮件地址、GUID、UPN 等用户属性中进行搜索。最关键的一点是,该参数支持以通配符开始的字符串,可以在查找同名用户、避免命名冲突方面发挥作用。命令和输出结果如下:

EXO PS > Get-Mailbox -Identity "*31" |
>> Select-Object DisplayName,RecipientTypeDetails, MicrosoftOnlineServicesID

```
DisplayName  RecipientTypeDetails  MicrosoftOnlineServicesID
-----------  --------------------  -------------------------
user 31      UserMailbox           user31@office365ps.partner.onmschina.cn
```

（3）当企业中的用户管理管理员或者全局管理员协助我们，为新用户分配合适的许可证后，我们是无法在 Exchange 管理中心的用户邮箱列表中，确认许可证的添加工作是否已经完成，并查看所有用户的许可证分配情况的。我们只能登录 Office 365 的管理中心，并在活跃用户列表中的状态列进行查看。而使用 PowerShell 命令，我们无须使用用户管理管理员或全局管理员的用户凭据重新登录，即可在同一个 PowerShell 的控制台窗口，查询新用户许可证的分配情况。命令和输出结果如下：

EXO PS > Get-Mailbox -Identity "user*" | Select-Object Displayname,SKUAssigned

```
DisplayName SKUAssigned
----------- -----------
user 01     True
user 02     True
user 21     True
user 31     True
user 32
user 33     True
user 34     True
user 35     True
user 36
```

在结果列表中，user 32 和 user 36 在 SKUAssigned 列的属性值为空，即处于未被分配许可证的状态。为了区分已分配许可证和未分配许可证的两类用户，我们可以进一步调整显示格式，将结果列表以分组的方式进行显示。命令和输出结果如下：

EXO PS > Get-Mailbox -Identity "user*" | Group-Object -Property SKUAssigned

```
Count Name                    Group
----- ----                    -----
    7 True                    {user 01, user 02, user 21, user 31...}
    2                         {user 32, user 36}
```

使用 Group-Object 命令，可以将查询结果依据 SKUAssigned 的状态显示成一个分组表格，并统计每个分组中成员的数量。类似地，我们可以借助 PowerShell 命令，查询在 Exchange 管理中心的用户邮箱列表中无法查询到的用户邮箱创建时间。命令和输出结果如下：

EXO PS > Get-Mailbox -Identity "user*" |
>> Select-Object DisplayName,WhenMailbox Created

```
DisplayName WhenMailboxCreated
----------- ------------------
user 01     7/27/2018 4:50:25 PM
user 02     10/7/2018 1:03:09 PM
user 21     10/7/2018 10:56:08 AM
user 31     10/20/2018 8:55:46 AM
user 32     10/20/2018 1:04:37 PM
user 33     10/21/2018 8:36:57 AM
```

```
user 34      10/21/2018 8:37:02 AM
user 35      10/21/2018 8:37:04 AM
user 36      10/21/2018 11:15:34 PM
```

> **注意**
> 如果企业的规模比较大，符合查询条件的记录数量大于 1000 条，且想返回全部的查询结果，可以使用 ResultSize 参数，并使用 Unlimited 这个属性值，也可以将该属性值设置为任意一个大于 1000，且小于 4294967296（2 的 32 次方）的整数。命令如下：

```
EXO PS > Get-Mailbox -ResultSize Unlimited
```

3. 用户邮箱属性的修改

（1）在 Exchange 管理中心中单击左侧导航栏中的"收件人"按钮，然后在右侧详细内容窗格中单击"邮箱"标签按钮，接着在用户邮箱列表中单击要修改邮箱属性用户的显示名称，并在页面右侧弹出的用户邮箱属性对话框中，对存档、OWA 等属性进行修改，如图 3-10 所示。

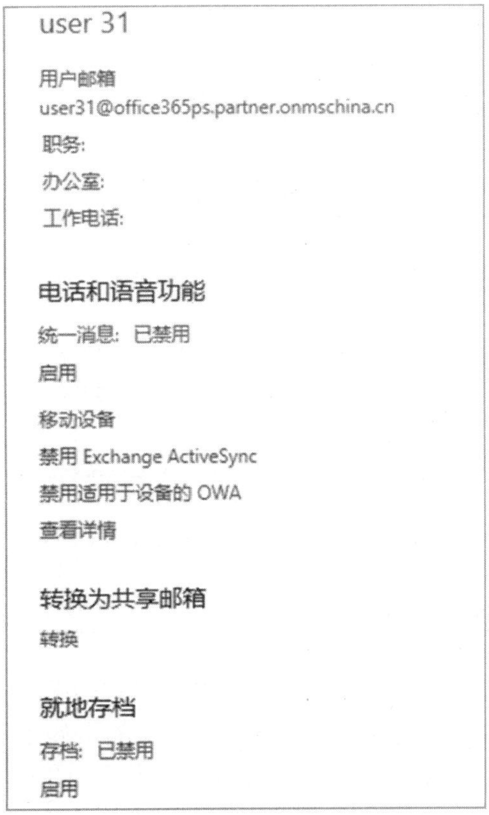

图 3-10

通过 PowerShell 命令设置用户邮箱的存档、OWA 等属性时，会涉及使用 Set-Mailbox、Set-User、Set-CASMailbox 和 Enable-Mailbox 等多条命令，具体的内容请参考表 3-1。

表 3-1

设置项目	PowerShell 命令实例
职务 工作电话 办公室	查看： EXO PS > Get-User -Identity "user31*" | Select-Object DisplayName,Title,Phone,Office 设置： EXO PS > Set-User -Identity "user31*" -Title "Director" -Phone "13612345678" -Office "306"
统一消息	需要结合"统一消息"进行设置。
Exchange ActiveSync	查看： EXO PS > Get-CASMailbox -Identity "user31*" | Select-Object DisplayName,ActiveSyncEnabled 关闭： EXO PS > Set-CASMailbox -Identity "user31*" -ActiveSyncEnabled $Ture 开启： EXO PS > Set-CASMailbox -Identity "user31*" -ActiveSyncEnabled $False
适用于设备的 OWA	查看： EXO PS > Get-CASMailbox -Identity "user31*" | Select-Object DisplayName,OWAforDevicesEnabled 开启： EXO PS > Set-CASMailbox -Identity "user31*" -OWAforDevicesEnabled $Ture 关闭： EXO PS > Set-CASMailbox -Identity "user31*" -OWAforDevicesEnabled $False
转换为共享邮箱	该内容将在本章后面的部分进行介绍。
就地存档	查看： EXO PS > Get-Mailbox -Identity "user31*" | Select-Object DisplayName,ArchiveStatus,ArchiveState 开启： EXO PS > Enable-Mailbox -Identity "user31@office365ps.partner.onmschina.cn" – Archive 关闭： EXO PS > Disable-Mailbox -Identity "user31@office365ps.partner.onmschina.cn" – Archive
就地保留	需要结合"合规性管理\就地电子数据展示和保留"进行设置。
Outlook 网页版	查看： EXO PS > Get-CASMailbox -Identity "user31*" | Select-Object DisplayName,OWAEnabled 开启： EXO PS > Set-CASMailbox -Identity "user31*" -OWAEnabled $Ture 关闭： EXO PS > Set-CASMailbox -Identity "user31*" -OWAEnabled $False

另外，我们可以使用之前介绍过的方法，将用户的标识和对应的需要为该用户设置的属性值存入 CSV 或文本文件，利用 foreach 循环来批量设置或修改用户邮箱的属性。

（2）在 Exchange 管理中心的用户邮箱列表中单击某个用户的显示名称，并单击菜单栏中的笔形"编辑"按钮，在弹出的"编辑用户邮箱"对话框中便可以查看或修改"邮箱使用情况""成员所属""邮箱委托"等邮箱属性。

在这些属性中，邮箱配额是 Exchange Online 管理员经常会关注到的属性之一。在"编辑用户邮箱"对话框左侧的导航栏中单击"邮箱使用情况"按钮，然后可以在右侧详细内容窗格中查看该用户邮箱配额的已用量，但是，仅可以查看配额的已用量，无法查看配额上限等其他信息，并且无法进行修改，如图 3-11 所示。

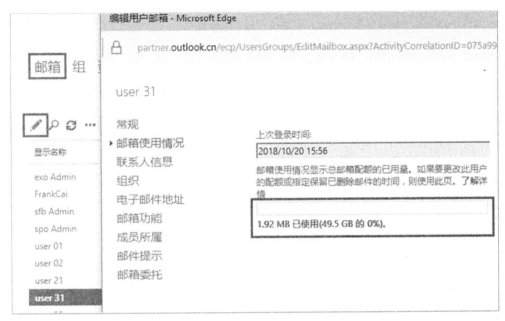

图 3-11

而我们使用下面的 PowerShell 命令可以列出有关邮箱配额的所有项目。命令和输出结果如下：

```
EXO PS > Get-Mailbox -Identity "user31*" | Format-List *Quota*

ProhibitSendQuota              : 49.5 GB (53,150,220,288 bytes)
ProhibitSendReceiveQuota       : 50 GB (53,687,091,200 bytes)
RecoverableItemsQuota          : 30 GB (32,212,254,720 bytes)
RecoverableItemsWarningQuota   : 20 GB (21,474,836,480 bytes)
CalendarLoggingQuota           : 6 GB (6,442,450,944 bytes)
UseDatabaseQuotaDefaults       : False
IssueWarningQuota              : 49 GB (52,613,349,376 bytes)
RulesQuota                     : 256 KB (262,144 bytes)
ArchiveQuota                   : 50 GB (53,687,091,200 bytes)
ArchiveWarningQuota            : 45 GB (48,318,382,080 bytes)
```

结果中有多个配额的阈值，而且配额的上限是由用户使用许可证的种类决定的。例如，使用 "Office 365 商业高级版" 许可证的用户，当邮箱容量达到 49GB 的时候，用户就会收到邮箱容量接近最大值的警告；当邮箱容量达到 49.5GB 的时候，就不允许再发送邮件了；当邮箱容量达到 50GB 容量的时候，收发邮件功能都将被限制。由于 user 31 开启了"就地存档"功能，存档邮箱的配额也同时进行了显示，用户存档邮箱同样也有 50GB 的配额上限，而当存档邮箱达到 45GB 容量的时候，用户就会收到存档邮箱容量接近最大值的告警。另外，我们还可以尝试使用下面的命令修改用户邮箱的配额。命令和输出结果如下：

EXO PS > Set-Mailbox -Identity "user31*" -ProhibitSendReceiveQuota 60GB

```
The operation on mailbox "user 31" failed because it's out of the current user's
write scope. The value of properties 'ProhibitSendReceiveQuota' exceeds the maximum
allowed for user 'user 31' with license 'BPOS_S_Standard'.
    + CategoryInfo          : InvalidOperation: (user 31:ADObjectId)
[Set-Mailbox], InvalidOperationException
    + FullyQualifiedErrorId                                            :
[Server=BJXPR01MB230,RequestId=e04d44e0-141b-40ba-ac71-f225d88a33da,TimeStamp
=10/23/2018  8:14:16  AM]  [FailureCategory=Cmdlet-InvalidOperationException]
2CD70E0D,Microsoft.Exchange.Management.RecipientTasks.SetMailbox
    + PSComputerName        : partner.outlook.cn
```

由于邮箱存储配额的上限是由用户使用许可证的种类决定的,而"Office 365 商业高级版"邮箱存储配额的最大值就是 50GB,所以会出现上面的报错。另外,并非邮箱存储配额的数值设定的比上限小,修改就会成功,请看下面的命令和输出结果:

EXO PS > Set-Mailbox -Identity "user31*" -ProhibitSendReceiveQuota 30GB `
>> -ProhibitSend Quota 35GB -IssueWarningQuota 36GB

```
The value of property 'IssueWarningQuota' must be less than or equal to that
of property 'ProhibitSendQuota'. IssueWarningQuota:
 '36 GB (38,654,705,664 bytes)', ProhibitSendQuota: '35 GB (37,580,963,840
bytes)'.
    + CategoryInfo          : NotSpecified: (user 31:ADObjectId) [Set-Mailbox],
DataValidationException
    + FullyQualifiedErrorId                                            :
[Server=BJXPR01MB230,RequestId=a7ec35bd-5b11-4f76-94ce-a71876283da1,TimeStamp
=10/23/2018  8:25:59  AM]  [FailureCategory=Cmdlet-DataValidationException]
248FE9DC,Microsoft.Exchange.Management.RecipientTasks.SetMailbox
    + PSComputerName        : partner.outlook.cn

The value of property 'ProhibitSendQuota' must be less than or equal to that
of property 'ProhibitSendReceiveQuota'.
 ProhibitSendQuota: '35 GB (37,580,963,840 bytes)', ProhibitSendReceiveQuota:
'30 GB (32,212,254,720 bytes)'.
    + CategoryInfo          : NotSpecified: (user 31:ADObjectId) [Set-Mailbox],
DataValidationException
    + FullyQualifiedErrorId                                            :
[Server=BJXPR01MB230,RequestId=a7ec35bd-5b11-4f76-94ce-a71876283da1,TimeStamp
=10/23/2018  8:26:00  AM]  [FailureCategory=Cmdlet-DataValidationException]
F003E380,Microsoft.Exchange.Management.RecipientTasks.SetMailbox
    + PSComputerName        : partner.outlook.cn
```

虽然我们可以将邮箱存储配额的数值设定得比最大值小,但是同样需要满足不同存储配额数值之间的依赖关系。IssueWarningQuota 应当小于等于 ProhibitSendQuota;ProhibitSendQuota 应当小于等于 ProhibitSendReceiveQuota;而 ProhibitSendReceiveQuota 应当小于等于该许可证所允许的最大邮箱配额。命令如下：

```
EXO PS > Set-Mailbox -Identity "user31*" -ProhibitSendReceiveQuota 30GB `
>> -ProhibitSend Quota 29.5GB -IssueWarningQuota 29GB
```

4. 用户邮箱的删除与恢复

一般公司都会有规定，人员离职的时候，该人员的邮箱是要被禁用或删除的，但是，在 Office 365 的 Exchange 管理中心或 Office 365 管理中心中，我们无法找到包含禁用邮箱或删除邮箱按钮的菜单或对话框。进行禁用邮箱或删除邮箱的操作，在本地部署的 Exchange 服务器（例如在 Exchange 2016）上，与在 Office 365 的 Exchange Online 上的具体步骤也有所不同。

（1）禁用 Exchange Online 中的邮箱。

在 Exchange Online 中移除之前给用户分配的许可证，或仅移除之前给用户分配的 Exchange Online 许可证，即可达到禁用该用户邮箱的目的。移除许可证的操作同样需要全局管理员或用户管理管理员的用户凭据。命令和输出结果如下：

```
O365 PS > Get-MsolUser -UserPrincipalName "user37@office365ps.partner.onmschina.cn"
```

```
UserPrincipalName                          DisplayName isLicensed
-----------------                          ----------- ----------
user37@office365ps.partner.onmschina.cn user 37        True
```

```
O365 PS > Set-MsolUserLicense `
>> -UserPrincipalName "user37@office365ps.partner.onmschina.cn" `
>> -RemoveLicenses "office365ps:O365_BUSINESS_PREMIUM"
EXO PS > Get-Mailbox -Identity "user37@office365ps.partner.onmschina.cn"
```

```
The operation couldn't be performed because object
'user37@office365ps.partner.onmschina.cn' couldn't be found
on 'SHAPR01A001DC01.CHNPR01A001.prod.partner.outlook.cn'.
    + CategoryInfo          : NotSpecified: (:) [Get-Mailbox],
ManagementObjectNotFoundException
    + FullyQualifiedErrorId :
[Server=BJXPR01MB230,RequestId=50afac20-431a-43dc-9058-4f7be15cfcf4,TimeStamp
=10/25/2018 8:41:25 AM]
[FailureCategory=Cmdlet-ManagementObjectNotFoundException]
4BC505E0,Microsoft.Exchange.Management.RecipientTasks.GetMailbox
    + PSComputerName        : partner.outlook.cn
```

移除用户 user 37 的"Office 365 商业高级版"的许可证并等待系统处理完毕后，当我们再次运行 Get-Mailbox 命令进行查询的时候，就会发现该用户的邮箱已经被禁用了。另外，由于用户 user 37 的邮箱不在许可证的缓冲期内了，所以给邮件地址 user37@office365ps.partner.onmschina.cn 发送邮件，会收到退信通知，如图 3-12 所示。

```
Delivery has failed to these recipients or groups:

user 37
This message was rejected by the recipient email system. Please check the recipient's
email address and try resending this message, or contact the recipient directly.
```

图 3-12

但是，用户 user 37 的 Office 365 账户依然存在，且状态正常。命令和输出结果如下：

O365 PS > Get-MsolUser -UserPrincipalName "user37@office365ps.partner.onmschina.cn"

```
UserPrincipalName                          DisplayName isLicensed
-----------------                          ----------- ----------
user37@office365ps.partner.onmschina.cn    user 37     False
```

禁用用户 user 37 的邮箱后，在 30 天的保留期内，如果我们重新为用户 user 37 分配"Office 365 商业高级版"的许可证，该邮箱的状态就可以恢复正常，且用户邮箱内已经接收的邮件也不会受到影响。命令和输出结果如下：

O365 PS > Set-MsolUserLicense `
>> -UserPrincipalName "user37@office365ps.partner.onmschina.cn" `
>> -AddLicenses "office365ps:O365_BUSINESS_PREMIUM"
EXO PS > Get-Mailbox -Identity "user37@office365ps.partner.onmschina.cn"

```
Name              Alias            Database              ProhibitSendQuota
ExternalDirectoryObjectId
----              -----            --------              -----------------
-------------------------
user37            user37           CHNPR01DG027-db035    49.5 GB
(53,150,2... 73263bec-8ef2-4878-8dd7-08ebb00c...
```

注意

在本地部署的 Exchange 服务器，例如 Exchange 2016 上，我们可以使用 Disable-Mailbox 来禁用邮箱，但是在 Office 365 的 Exchange Online 上，该命令只能用于禁用归档邮箱。命令和输出结果如下：

EXO PS > Get-Mailbox -Archive

```
Name              Alias            Database              ProhibitSendQuota
ExternalDirectoryObjectId
----              -----            --------              -----------------
-------------------------
user36            user36           CHNPR01DG027-db056    49.5 GB
(53,150,2... 2201bfa8-c191-4e6e-bcd2-5c68946a...
```

第 3 章 使用 PowerShell 管理 Office 365 的 Exchange Online

EXO PS > Disable-Mailbox -Archive -Identity "user36@office365ps.partner.onmschina.cn"

```
Confirm
Are you sure you want to perform this action?
Disabling the archive for "user36@office365ps.partner.onmschina.cn" will
remove the archive for this user and mark it in the database for removal.
[Y] Yes  [A] Yes to All  [N] No  [L] No to All  [?] Help (default is "Y"): Y
```

EXO PS > Get-Mailbox -Archive

我们在运行 Disable-Mailbox 命令之前可以查询到用户 user 36 的归档邮箱，而运行完 Disable-Mailbox 命令后，当我们再次运行 Get-Mailbox 命令时，却无法再次查看到用户 user 36 的归档邮箱，因为该归档邮箱的状态已经变为禁用了。

如果我们使用 Disable-Mailbox 命令来禁用普通的用户邮箱，我们会得到需要使用 PermanentlyDisable 参数的提示，但是，即使我们按照提示使用了 PermanentlyDisable 参数，命令也无法成功执行，还是会得到下面的提示。命令和输出结果如下：

EXO PS > Disable-Mailbox -Identity "user35@office365ps.partner.onmschina.cn" `
\>> -PermanentlyDisable

```
Cannot Disable-Mailbox for 'user 35' because this user has a valid license.
    + CategoryInfo          : NotSpecified: (user 35:ADObjectId)
[Disable-Mailbox], RecipientTaskException
    + FullyQualifiedErrorId                                         :
[Server=BJXPR01MB230,RequestId=33e55e0f-40dd-4647-b0cb-18d9a8635520,TimeStamp
=5/27/2019   11:39:43   AM] [FailureCategory=Cmdlet-RecipientTaskException]
BC0AF995,Microsoft.Exchange.Management.RecipientTasks.DisableMailbox
    + PSComputerName        : partner.outlook.cn
```

即使我们接着移除了用户 user 35 的"Office 365 商业高级版"的许可证，在 Exchange Online 上，仍然无法使用 Disable-Mailbox 命令来禁用用户邮箱。命令和输出结果如下：

EXO PS > Disable-Mailbox -Identity "user35@office365ps.partner.onmschina.cn" `
\>> -PermanentlyDisable

```
The operation couldn't be performed because object
'user35@office365ps.partner.onmschina.cn' couldn't be found on
'SH0PR01A01DC002.CHNPR01A001.prod.partner.outlook.cn'.
    + CategoryInfo          : NotSpecified: (:) [Disable-Mailbox],
ManagementObjectNotFoundException
    + FullyQualifiedErrorId  :  [Server=BJXPR01MB230,RequestId=
d606fc01-5e7c-44ba-b7b9-1340303a1058,TimeStamp=5/27/2019   11:34:41   AM]
[FailureCategory=Cmdlet-ManagementObjectNotFoundException]
9ED41354,Microsoft.Exchange.Management.RecipientTasks.DisableMailbox
    + PSComputerName        : partner.outlook.cn
```

有个例外，如果使用 New-Mailbox 命令创建了用户邮箱，并隐式地创建了用户账户，在没有给该用户分配许可证的前提下，使用 Disable-Mailbox 命令可以在 Exchange Online 上禁用该用户的邮箱。禁用邮箱的操作将该邮箱的 Exchange 属性删除，该邮箱也会在操作完成后立即被标记为分离的状态。命令和输出结果如下：

```
EXO PS > Disable-Mailbox -Identity "user33@office365ps.partner.onmschina.cn" `
>> -PermanentlyDisable
```

```
Confirm
Are you sure you want to perform this action?
Disabling mailbox "user33@office365ps.partner.onmschina.cn" will remove the
Exchange properties from the Active Directory user object and mark the mailbox
 in the database for removal. If the mailbox has an archive or remote archive,
 the archive will also be marked for removal. In the case of remote archives,
 this action is permanent. You can't reconnect this user to the remote archive
 again.
[Y] Yes  [A] Yes to All  [N] No  [L] No to All  [?] Help (default is "Y"): Y
```

该邮箱相关的用户账户并不会被清除，类型会变更为 User，我们可以使用 Get-User 命令进行查看。命令和输出结果如下：

```
EXO PS > Get-User -Identity "user33*" | Select-Object DisplayName,RecipientTypeDetails
```

```
DisplayName  RecipientTypeDetails
-----------  --------------------
user 33      User
```

如果稍后要恢复用户 user 33 的邮箱，并不需要使用 Enable-Mailbox 命令，而只需要为用户 user 33 分配"Office 365 商业高级版"的许可证，就可以完成恢复邮箱的工作了。命令和输出结果如下：

```
O365 PS > Set-MsolUser `
>> -UserPrincipalName "user33@office365ps.partner.onmschina.cn" `
>> -UsageLocation "CN"
O365 PS > Set-MsolUserLicense `
>> -UserPrincipalName "user33@office365ps.partner.onmschina.cn" `
>> -AddLicenses "office365ps:O365_BUSINESS_PREMIUM"
O365 PS > Get-MsolUser `
>> -UserPrincipalName "user33@office365ps.partner.onmschina.cn" |
>> Select-Object Licenses
```

```
Licenses
--------
{office365ps:O365_BUSINESS_PREMIUM}
```

最后，我们可以确认用户 user 33 的邮箱已经被恢复了，且用户的状态已经变成 UserMailbox 了。命令和输出结果如下：

EXO PS > Get-Mailbox -Identity "user33*"

```
Name                                   Alias                   Database
ProhibitSendQuota    ExternalDirectoryObjectId
----                                   -----                   --------
----------------     -------------------------
user33                                 user33                  CHNPR01DG029-db055             49.5 GB
(53,150,2... f9faffef-f69f-4e2a-9e21-51bb9a39...
```

EXO PS > Get-User -Identity "user33*" | Select-Object DisplayName,RecipientTypeDetails

```
DisplayName  RecipientTypeDetails
-----------  --------------------
user 33      UserMailbox
```

（2）删除 Exchange Online 中的邮箱。

在本地部署的 Exchange 服务器，例如在 Exchange 2016 上，Exchange 的管理员可以在 Exchange 管理中心使用删除邮箱按钮来删除邮箱。然而，在 Office 365 的 Exchange 管理中心或者 Office 365 管理中心中却没有删除邮箱的按钮。为了可以在 Office 365 中进行删除邮箱的操作，Exchange Online 管理员可以使用 Remove-Mailbox 这个 PowerShell 命令。命令和输出结果如下：

EXO PS > Remove-Mailbox -Identity "user34@office365ps.partner.onmschina.cn"

```
Confirm
Are you sure you want to perform this action?
Removing the mailbox "user34@office365ps.partner.onmschina.cn" will mark the
mailbox and the archive, if present, for deletion. The associated Windows Live
 ID "user34@office365ps.partner.onmschina.cn" will also be deleted and will
not be available for any other Windows Live service.
[Y] Yes  [A] Yes to All  [N] No  [L] No to All  [?] Help (default is "Y"): Y
```

在本地部署的 Exchange 服务器，例如在 Exchange 2016 上，"删除邮箱"的操作同时会在活动目录中删除该用户。类似地，在 Office 365 的 Exchange Online 中，使用删除邮箱的 PowerShell 命令也会同时删除该用户在 Office 365 中的用户账户。如果使用 Get-User 命令再次查询该用户账户，会收到以下的报错：

EXO PS > Get-User -Identity "user34@office365ps.partner.onmschina.cn"

```
The operation couldn't be performed because object 'user34*' couldn't be found
on
  'SH2PR01A001DC06.CHNPR01A001.prod.partner.outlook.cn'.
    + CategoryInfo                    : NotSpecified: (:) [Get-User],
ManagementObjectNotFoundException
    + FullyQualifiedErrorId :
[Server=BJXPR01MB230,RequestId=f600d901-658e-4c93-8e84-42a079a8d108,TimeStamp
=10/24/2018 5:01:19 AM]
```

```
[FailureCategory=Cmdlet-ManagementObjectNotFoundException]
CB0371C,Microsoft.Exchange.Management.RecipientTasks.GetUser
    + PSComputerName        : partner.outlook.cn
```

虽然，我们已经无法使用 Get-User 命令查询用户 user 34 的信息了，但是，由于该用户刚刚被删除，在 Office 365 管理中心中单击左侧导航栏中的"用户"按钮，在弹出的扩展菜单中单击"已删除的用户"按钮，仍然可以在右侧的"已删除用户"列表中看到被删除的 user 34 用户，如图 3-13 所示。

图 3-13

类似地，Exchange Online 管理员可以使用 Connect-MsolService 命令连接 Office 365 的租户，并使用 PowerShell 命令在被删除的用户列表中查询到用户 user 34。命令和输出结果如下：

EXO PS > $cred = Get-Credential
EXO PS > Connect-MsolService -Credential $cred -AzureEnvironment AzureChinaCloud
EXO PS > Get-MsolUser -ReturnDeletedUsers

```
    UserPrincipalName
DisplayName           isLicensed
    -----------------
-----------           ----------
ExRemoved-c665f13d6e4f4c7d9f61494e836f1198@office365ps.partner.onmschina.cn
user 34               True
```

我们也可以使用下面的 Get-Mailbox 命令查询被删除的邮箱。不过要注意，这些邮箱不是被彻底地删除了，而是会在系统中保留 30 天，即这些邮箱被"软删除"了。对比之前提到的内容，在 Exchange Online 中被禁用邮箱是无法使用下面的命令查询到的。命令和输出结果如下：

EXO PS > Get-Mailbox -SoftDeletedMailbox

```
    Name              Alias         Database              ProhibitSendQuota
ExternalDirectoryObjectId
```

```
        ----                      -----                    --------                              ----------------
------------------------
        user 34                   user34                   CHNPR01DG028-db104                           49.5 GB
(53,150,2... bc9116f6-86ce-4c84-a368-90f18891...
```

被软删除的用户邮箱在系统保留期内，仍然可以被恢复，且恢复该邮箱的同时，该用户账户也同时会被恢复。命令和输出结果如下：

EXO PS > Undo-SoftDeletedMailbox `
>> -SoftDeletedObject "user34@office365ps.partner. onmschina.cn"

```
Name                                                       Alias                                 Database
ProhibitSendQuota    ExternalDirectoryObjectId
----                                                       -----                                 --------
-----------------    -------------------------
        user 34                   user34                   CHNPR01DG028-db104                           49.5 GB
(53,150,2... bc9116f6-86ce-4c84-a368-90f18891...
```

恢复邮箱的操作完成后，邮箱被删除前已经接收的邮件不会受到影响。另外，我们可以使用 Get-User 命令和 Get-Mailbox 命令，来确认用户 user 34 及其邮箱的状态。命令和输出结果如下：

EXO PS > Get-User -Identity "user34*" |
>> Select-Object DisplayName,SKUAssigned, RecipientTypeDetails

```
DisplayName SKUAssigned RecipientTypeDetails
----------- ----------- --------------------
user 34            True UserMailbox
```

EXO PS > Get-Mailbox -Identity "user34@office365ps.partner.onmschina.cn"

```
Name                                                       Alias                                 Database
ProhibitSendQuota    ExternalDirectoryObjectId
----                                                       -----                                 --------
-----------------    -------------------------
        user34                    user34                   CHNPR01DG040-db073                           49.5 GB
(53,150,2... 0eb32fa0-9ec8-42e8-8dfc-3da148a6...
```

（3）彻底删除 Exchange Online 中的邮箱。

虽然在 Exchange 管理中心的用户邮箱列表中，Exchange Online 管理员无法进行删除邮箱的操作，但是却可以使用 Remove-Mailbox 命令来删除用户邮箱，而且该命令还同时将用户移入本租户的已删除用户列表。要想彻底删除该用户的邮箱，还要求该邮箱不能是非活跃邮箱。

注意

非活跃邮箱是用户邮箱的一种状态。把邮箱设置成非活跃邮箱后，该邮箱便不再受系统删除保留期默认 30 天的限制了，而且我们还可以进一步修改非活跃邮箱被保留的时间。另外，设置非活跃邮箱还需要为租户购买 Office 365 Enterprise E3 或者 Office 365 Enterprise E5 的订阅，如果租户只有 Office 365 Enterprise E1 的订阅，需要额外添加邮箱"归档"许可证。例如，虽然用户 user 39 离职了，但是该用户邮箱中的来往邮件是与业务高度相关的，必须保留。而且后续还要让其他同事来访问该邮箱中的内容，或者把该邮箱中的信件合并到其他同事的邮箱中。如果我们没有将 user 39 的用户邮箱设置为非活跃邮箱，便进行了软删除的操作，在邮箱 30 天的保留期后，该邮箱就会被彻底地移除了。

彻底删除 Exchange Online 中的邮箱可以按照下述最佳操作步骤来进行。首先，对用户邮箱进行软删除的操作，而该操作也会同时删除对应的用户账户。命令和输出结果如下：

EXO PS > Remove-Mailbox -Identity "user38@office365ps.partner.onmschina.cn"
\>\> -Confirm:$false

然后，以全局管理员或账户管理员的用户凭据，使用 AAD Module 连接 Office 365 的租户，并将上述用户从租户的已删除用户的列表中彻底删除。命令和输出结果如下：

O365 PS > Get-MsolUser -ReturnDeletedUsers |
\>\> Where-Object {$_.DisplayName -match "user 38"} |
\>\> Remove-MsolUser -RemoveFromRecycleBin

```
Confirm
Continue with this operation?
[Y] Yes  [N] No  [S] Suspend  [?] Help (default is "Y"): Y
```

最后，我们就可以将该用户的邮箱彻底删除了。在进行用户邮箱的彻底删除操作后，这个被删除的邮箱就无法进行恢复了。命令和输出结果如下：

EXO PS > Remove-Mailbox -Identity "user38@office365ps.partner.onmschina.cn" `
\>\> -PermanentlyDelete

```
Confirm
Are you sure you want to perform this action?
The mailbox "user38@office365ps.partner.onmschina.cn" is about to be
permanently deleted. All mailbox content such as emails, contacts, etc. will
be irrecoverably deleted.
[Y] Yes  [A] Yes to All  [N] No  [L] No to All  [?] Help (default is "Y"): Y
```

如果不首先进行邮箱的"软删除"操作，或是不首先使用上述 Remove-MsolUser 命令将用户从已删除用户的列表中彻底删除，而是在"软删除"操作后直接使用 Remove-Mailbox 命令和 PermanentlyDelete 参数彻底删除邮箱，我们都会收到下面的报错信息。命令和输出结果如下：

EXO PS > Remove-Mailbox -Identity "user38@office365ps.partner.onmschina.cn" `
\>\> -PermanentlyDelete

```
Confirm
Are you sure you want to perform this action?
Removing the mailbox "user38@office365ps.partner.onmschina.cn" will mark the
mailbox and the archive, if present, for deletion. The associated Windows Live
 ID "user38@office365ps.partner.onmschina.cn" will also be deleted and will
not be available for any other Windows Live service.
 [Y] Yes  [A] Yes to All  [N] No  [L] No to All  [?] Help (default is "Y"): Y
Error on proxy command 'Remove-Mailbox
-Identity:'user38@office365ps.partner.onmschina.cn' -PermanentlyDelete:$True
-Confirm:$False' to server SHXPR01MB0640.CHNPR01.prod.partner.outlook.cn: Server
version 15.20.1900.0000, Proxy method PSWS:
Cmdlet error with following error message:
Microsoft.Exchange.Management.Tasks.RecipientTaskException: This  mailbox
cannot be permanently deleted since there is a user associated with this mailbox
in Azure Active Directory. You will first need to delete the user in Azure Active
Directory. Please refer to documentation for more details.
   at    Microsoft.Exchange.Configuration.Tasks.Task.ThrowError(Exception
exception, ErrorCategory errorCategory, Object target, String helpUrl)
   at    Microsoft.Exchange.Configuration.Tasks.Task.WriteError(Exception
exception, ErrorCategory category, Object target, Boolean reThrow)
   at
Microsoft.Exchange.Management.RecipientTasks.RemoveMailboxBase`2.InternalVali
date()
   at
Microsoft.Exchange.Management.RecipientTasks.RemoveMailbox.InternalValidate()
   at Microsoft.Exchange.Configuration.Tasks.Task.<ProcessRecord>b__93_1()
   at
Microsoft.Exchange.Configuration.Tasks.Task.InvokeRetryableFunc(String
funcName, Action func, Boolean
   terminatePipelineIfFailed).
[Server=BJXPR01MB230,RequestId=a1fe19c3-2b28-41fe-8e8e-ecb34dafa8b9,TimeStamp
=5/29/2019 8:26:02 AM] .
   + CategoryInfo            : NotSpecified: (:) [Remove-Mailbox],
CmdletProxyException
   +                           FullyQualifiedErrorId                        :
Microsoft.Exchange.Configuration.CmdletProxyException,Microsoft.Exchange.Mana
gement.RecipientTasks
    .RemoveMailbox
   + PSComputerName          : partner.outlook.cn
```

（4）禁用和删除用户邮箱的要点。

根据企业中不同的应用场景，我们可以采用禁用或删除用户邮箱的方法，来实现不同的需求。

第一种应用场景：我们要保留用户账户，而禁用该用户所使用的邮箱。我们可以通过删除该用户所拥有的许可证来禁用该邮箱，且被禁用的邮箱无法使用 Get-Mailbox 命令查询到。而在禁用邮箱被删除前的保留期内，可以为该用户重新分配许可证来恢复邮箱。使用

New-Mailbox 命令新创建的用户邮箱及用户账户，在用户没有被分配许可证前，可以使用 Disable-Mailbox 命令来禁用该邮箱。

第二种应用场景：我们要删除用户邮箱和用户账户。我们可以先使用 Remove-Mailbox 命令软删除该邮箱，被软删除的邮箱可以使用 Get-Mailbox 命令进行查询，并且该用户也会同时出现在已删除用户的列表中。类似地，在该邮箱被彻底删除前的邮箱保留期内，可以使用 Undo-SoftDeletedMailbox 命令来恢复邮箱和相关联的用户账户，或者也可以从已删除用户的列表中彻底删除该用户后，最终彻底删除该邮箱。

第三种应用场景：我们要在很长的一段时间内保留特定的邮箱，而该邮箱也不能受系统保留周期的影响。我们可以将该邮箱设置为非活跃邮箱，但前提是企业必须拥有 Office 365 Enterprise E3 或 Office 365 Enterprise E5 的订阅。

另外，我们在进行上述禁用邮箱或者删除邮箱的操作时，给 Identity 参数赋值时不要使用 "user3*" 这种包含通配符的字符串来作为邮箱的标识，而要使用 user31@office365ps.partner.onmschina.cn 作为邮箱的标识。因为在进行禁用邮箱或者删除邮箱的操作时，如果该通配符可以与多个用户标识匹配，将会出现下面的报错。命令和输出结果如下：

EXO PS > Remove-Mailbox -Identity "user3*"

```
The operation couldn't be performed because 'user3*' matches multiple entries.
    + CategoryInfo          : NotSpecified: (:) [Remove-Mailbox],
ManagementObjectAmbiguousException
    + FullyQualifiedErrorId :
[Server=BJXPR01MB230,RequestId=50afac20-431a-43dc-9058-4f7be15cfcf4,TimeStamp
=10/26/2018 4:46:22 AM]
[FailureCategory=Cmdlet-ManagementObjectAmbiguousException]
98CBBAEE,Microsoft.Exchange.Management.RecipientTasks.RemoveMailbox
    + PSComputerName        : partner.outlook.cn
```

3.2.2 如何管理组

在本书的第 2 章中，我们介绍了安全组的管理。接下来我们就来了解如何管理其他几种类型的组。如果我们仅使用 Microsoft Azure Active Directory 模块连接 Office 365 的租户，是无法使用 PowerShell 命令管理其他这几种类型的组的。进行管理工作的前提条件是创建一个与 Exchange Online 的连接。

1. 通讯组列表和启用邮件的安全组的管理

（1）通讯组列表和启用邮件的安全组的创建。

顾名思义，通讯组列表就是为了给该组中的成员发送邮件消息而存在的。在 Exchange Online 管理中心单击左侧导航栏中的"收件人"按钮，然后在右侧详细内容窗格上方单击"组"标签按钮，接着单击组列表菜单中的倒三角形状的"新建"按钮，并在弹出的扩展菜单中单击"通讯组列表"按钮，来进行添加，如图 3-14 所示。

第 3 章 使用 PowerShell 管理 Office 365 的 Exchange Online

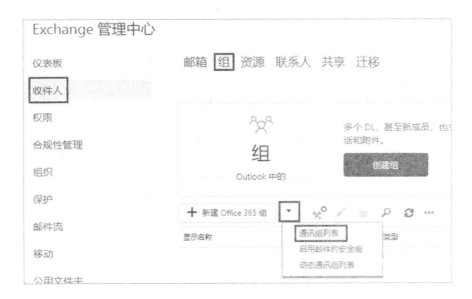

图 3-14

在"新建通讯组列表"对话框中，必须填写的项目包括"显示名称""别名"等，即那些带星号的项目，如图 3-15 所示。

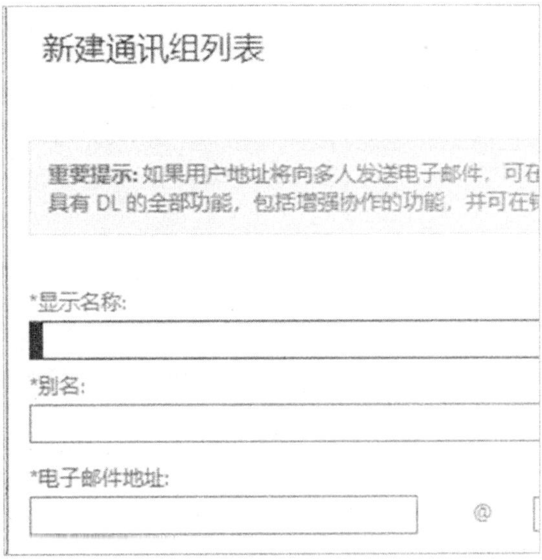

图 3-15

而使用 New-DistributionGroup 命令来创建通讯组列表，我们可以仅添加 Name 这个参数。命令和输出结果如下：

EXO PS > New-DistributionGroup -Name "disSales"

```
New! Office 365 Groups are the next generation of distribution lists.
Groups give teams shared tools for collaborating using email, files, a calendar,
```

```
and more.
  You can start right away using the New-UnifiedGroup cmdlet.

  Name       DisplayName  GroupType  PrimarySmtpAddress
  ----       -----------  ---------  ------------------
  disSales   disSales     Universal  disSales@office365ps.partner.onmschina.cn
```

如果在 New-DistributionGroup 命令中，我们只添加了 Name 参数，显示名称和别名的属性值也会使用 Name 参数的值来设定。另外，如果命令中没有添加 ManagedBy 参数，即通讯组列表的所有者这个参数，执行 New-DistributionGroup 命令的用户，就默认会成为该新建组的所有者。在"新建通讯组列表"对话框中的"选择是否需要所有者批准才能加入组"和"选择组是否允许自由退出"这两个属性上，如果没有在 PowerShell 命令中添加相应的参数，将会使用"通讯组列表"的默认设置，即"通讯组列表"不允许用户自己加入该通讯组列表，而允许用户自行离开该通讯组列表。

如果要添加启用邮件的安全组，我们可以在组列表菜单栏中单击倒三角形状的"新建"按钮，然后在弹出的扩展菜单中单击"启用邮件的安全组"按钮，请参见图 3-14。

注意

创建启用邮件的安全组并不是用 New-MsolGroup 命令，而是用 New-DistributionGroup 命令。

在 Exchange Online 中，所有可以支持邮件功能的组类型，都可以定义为通讯组或者分发组，而不论该种类型的组是否带有安全属性。所以将启用邮件的安全组称作"有安全属性的通讯组"似乎更为贴切，也能更好地说明这种组类型的功能特性。创建启用邮件的安全组的命令和输出结果如下：

EXO PS > New-DistributionGroup -Name "sdiSales" -Type "Security"

```
New! Office 365 Groups are the next generation of distribution lists.
Groups give teams shared tools for collaborating using email, files, a calendar,
and more. You can start right away using the New-UnifiedGroup cmdlet.

Name       DisplayName  GroupType                     PrimarySmtpAddress
----       -----------  ---------                     ------------------
sdiSales   sdiSales     Universal, SecurityEnabled    sdiSales@office365ps.partner.
onmschina.cn
```

命令返回结果中的 GroupType 属性，通讯组列表该属性的值是 Universal，而启用邮件的安全组该属性的值是 Universal, SecurityEnabled。表明启用邮件的安全组既可以用于向该组中的成员分发邮件消息，也可以作为租户中资源的权限授予对象。另外，启用邮件的安全组在组成员加入、退出属性上的默认设置，也和通讯组列表有些区别，前者由于会被授予资源访问的权限，不允许组成员的自行加入与自行离开，组成员的加入与离开需要由组的所有者决定，而后者在成员的加入和退出上则更为灵活。启用邮件的安全组和通讯组列表在加入组与退出组设置项上的差异请参考表 3-2。

表 3-2

操　作	设　置　项	通讯组列表	启用邮件的安全组
加入组	开放（加入无须经组的所有者批准）	默认设置	不可以设置
	关闭（自动拒绝所有加入请求）	可以设置	默认设置
	所有者审批（所有请求必须由所有者批准）	可以设置	不可以设置*
退出组	开放（退出无须经组的所有者批准）	默认设置	不可以设置
	关闭（自动拒绝所有退出请求）	可以设置	默认设置

*虽然启用邮件的安全组可以设置为"成员需要审批加入"，但是审批请求是不会发送给该组的管理员的，所以该设置并不起作用。

在可以被授予访问权限的这点上，启用邮件的安全组和安全组是类似的，而安全组没有邮件地址，不能用来向组中的成员分发邮件消息。

（2）通讯组列表和启用邮件的安全组的组成员管理。

Exchange Online 的管理员一般是组的创建者，即所有者，拥有管理组成员的权限。并且，通讯组列表和启用邮件的安全组在添加、删除成员上的操作步骤是类似的。

我们可以在新建组时为该组添加、删除组成员，即在"新建通讯组列表"或"新建启用邮件的安全组"对话框中的"成员"列表项目处单击"+"按钮，然后在弹出的"选择成员"对话框中，选择并添加该组的成员；或者选中"成员"列表中的成员对象，然后单击"-"按钮进行删除。也可以在新组创建完成后，在组列表中单击特定组的显示名称，接着在组列表菜单栏中单击笔形的"编辑"按钮。在弹出的"编辑通讯组列表"或"编辑启用邮件的安全组"对话框中单击左侧导航栏中的"成员身份"按钮，然后在右侧的成员列表的菜单栏中单击"+"或"-"按钮来添加或删除组成员，如图 3-16 所示。

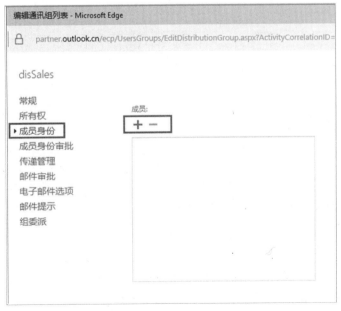

图 3-16

类似地，使用 New-DistributionGroup 命令，可以在新创建通讯组列表或启用邮件的安全组的时候就添加组成员。要注意的是，各个组成员间是用逗号进行分隔的，且组成员的名称不能添加引号，要不然会出现 Couldn't find object 的错误。命令和输出结果如下：

EXO PS > New-DistributionGroup -Name "sdiHR" -Type "Security" -Members `
>> user33@office365ps.partner.onmschina.cn,user34@office365ps.partner.onmschina.cn

```
New! Office 365 Groups are the next generation of distribution lists.
Groups give teams shared tools for collaborating using email, files, a calendar,
and more. You can start right away using the New-UnifiedGroup cmdlet.

Name   DisplayName  GroupType  PrimarySmtpAddress
----   -----------  ---------  ------------------
sdiHR  sdiHR        Universal  sdiHR@office365ps.partner.onmschina.cn
```

新建组或编辑组属性，并通过"选择成员"对话框向组成员列表添加成员时，我们在备选成员列表中查询到的只是备选对象的显示名称和电子邮件地址。而要想通过其他条件查找成员，可以在备选成员列表的菜单栏中单击"…"按钮，在弹出的扩展菜单中单击"高级搜索"按钮，并在"高级搜索"对话框中添加筛选条件。比如我们要添加Sales部门的用户到sdiSales组，可以选择"*部门："左侧的复选框，并在该属性右侧的文本框中键入Sales，如图3-17所示。

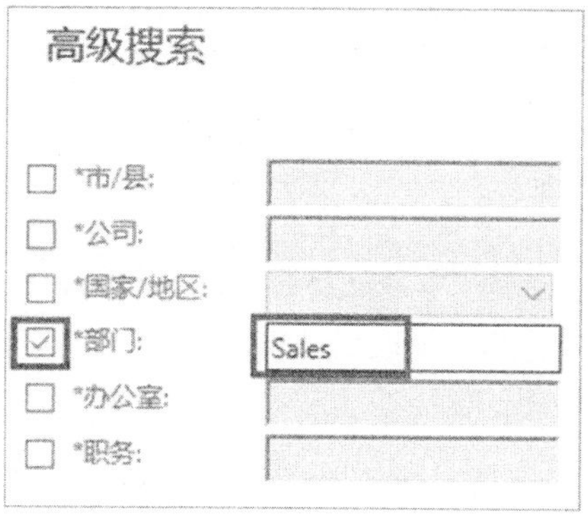

图 3-17

而使用 PowerShell 命令将整合以上的这些步骤，高效地完成 Sales 组成员的添加工作。首先我们通过 Get-User 命令查找存在用户邮箱且部门属性是 Sales 的用户，并把这些用户的标识名称存入数组$salesMember；然后将该数组中的用户标识名称利用循环，逐一赋值给 Add-DistributionGroupMember 命令的 Member 参数，完成组成员的添加。命令如下：

EXO PS > $salesMember = @((Get-User |
>> Where-Object {($_.RecipientTypeDetails -match "UserMailbox") `
-and ($_.Department -match "Sales")}).UserPrincipalName)
EXO PS > $salesMember | ForEach-Object {
>> Add-DistributionGroupMember -Identity "sdiSales" -Member $_ }

删除通讯组列表或启用邮件的安全组中的成员可以使用下面的 PowerShell 命令。命令和输出结果如下：

EXO PS > Remove-DistributionGroupMember -Identity "sdiHR" `
\>\> -Member user33@office365ps.partner.onmschina.cn

```
Confirm
Are you sure you want to perform this action?
Removing distribution group member "user33@office365ps.partner.onmschina.cn"
from distribution group "sdiHR".
[Y] Yes  [A] Yes to All  [N] No  [L] No to All  [?] Help (default is "Y"): Y
```

要想替换通讯组列表或启用邮件的安全组中多数的成员，可以首先删除该组内所有的成员，然后添加要加入的新成员。删除特定组中所有成员的命令如下：

EXO PS > Get-DistributionGroupMember -Identity "sdiHR" | ForEach-Object {
\>\> Remove-DistributionGroupMember -Identity "sdiHR" `
\>\> -Member $_.WindowsLiveID -Confirm:$false}

其实，我们还可以使用 Update-DistributionGroupMember 命令，直接将原有的组成员列表替换为新提供的组成员列表。命令和输出结果如下：

EXO PS > Update-DistributionGroupMember -Identity "sdiHR" -Members `
\>\> user21@office365ps.partner.onmschina.cn,user31@office365ps.partner.onmschina.cn

```
Confirm
Are you sure you want to perform this action?
Updating the membership list on group "sdiHR".
[Y] Yes  [A] Yes to All  [N] No  [L] No to All  [?] Help (default is "Y"): Y
```

最后，通过使用 Get-DistributionGroupMember 命令，我们可以确认启用邮件的安全组 sdiHR 的组成员列表中，目前只包含用户 user 21 和 user 31。命令和输出结果如下：

EXO PS > Get-DistributionGroupMember -Identity "sdiHR"

```
Name     RecipientType
----     -------------
user21   UserMailbox
user31   UserMailbox
```

注意

由于为租户中的启用邮件的安全组授予权限的时候，该组中没有安全属性的对象，例如联系人或通讯组列表等对象将无法获得这些被授予的权限。所以我们在给启用邮件的安全组添加组成员时，建议只添加有用户邮箱的用户，或其他启用邮件的安全组等带有安全属性的对象。

（3）通讯组列表和启用邮件的安全组的查看。

在 Exchange 管理中心的组列表中，默认不仅显示通讯组列表和启用邮件的安全组，还显示动态通讯组列表和 Office 365 组。在组列表菜单栏中，如果没有搜索文本框，可以单击放大镜形状的"搜索"按钮，使搜索文本框弹出；然后在搜索文本框中键入搜索关键字，并单击放大镜形状的"搜索组"按钮，即可查找特定的组，但是，该搜索功能依然使用 Anr 参数，所以搜索字符串不能以"*"开头。虽然单击菜单栏中的"..."按钮，并在弹出的扩展菜单中单击"高级搜索"按钮，可以在"高级搜索"对话框中指定更多的搜索条件，但是，搜索关键字同样不支持以"*"开头的字符串，且如果使用名称的部分作为搜索关键字，将可能导致搜索结果的异常。

虽然使用 Get-DistributionGroup 命令只能查询到通讯组列表和启用邮件的安全组这两种类型的组，但是，在命令执行的结果中会包括会议室的通讯组列表，在 Exchange 管理中心的组列表中则无法查看会议室的通讯组列表。命令和输出结果如下：

```
EXO PS > Get-DistributionGroup

Name              DisplayName        GroupType                   PrimarySmtpAddress
----              -----------        ---------                   ------------------
Cr3F                                 Cr3F                        Universal            Cr3F@office365ps.partner.onmschina.cn
disSales                             disSales                    Universal            disSales@office365ps.partner.onmschina.cn
sdiHR                                sdiHR                       Universal, SecurityEnabled            sdiHR@office365ps.partner.onmschina.cn
```

通过使用 Get-DistributionGroup 命令，并配合相应的参数，我们可以分别产生通讯组列表或启用邮件的安全组的结果集，可以从结果中排除会议室的通讯组列表，而且还可以使用以"*"开头的搜索字符串。命令和输出结果如下：

```
EXO PS > Get-DistributionGroup |
>> Where-Object {($_.GroupType -eq "Universal") `
>> -and ($_.RecipientTypeDetails -ne "RoomList")} |
>> Where-Object {$_.DisplayName -ilike "*Sale*"}

Name       DisplayName  GroupType   PrimarySmtpAddress
----       -----------  ---------   ------------------
disSales   disSales     Universal   disSales@office365ps.partner.onmschina.cn
```

（4）通讯组列表和启用邮件的安全组的删除。

在 Exchange 管理中心的组列表中单击要删除的通讯组列表或启用邮件的安全组的显示名称，然后单击组列表菜单栏中垃圾桶形状的"删除"按钮，并在进行确认后删除该组。相应地，我们可以使用 Remove-DistributionGroup 命令删除通讯组列表或启用邮件的安全组，必选的参数是即将被删除的组的名称。命令和输出结果如下：

```
EXO PS > Remove-DistributionGroup -Identity "sdiSales"

Confirm
Are you sure you want to perform this action?
```

```
Removing distribution group "sdiSales" will remove the Active Directory group
object.
[Y] Yes  [A] Yes to All  [N] No  [L] No to All  [?] Help (default is "Y"): Y
```

2. 动态通讯组列表的管理

通讯组列表或启用邮件的安全组的组成员列表的维护，一般要依赖于 Exchange Online 的管理员，伴随着组织规模的扩大，人员的频繁变动，尤其是人员部门的变更，会使组成员列表维护的工作量不断增加。例如用户从 HR 部门调到了 Sales 部门，如果使用的是通讯组列表，且组成员维护的规则被设定为拒绝所有成员的加入和退出请求，则 Exchange Online 管理员需要把该用户从原有的组成员列表中删除，然后添加到新的列表当中。然而，通过使用动态通讯组列表，上述组成员列表维护工作便会减少。

（1）虽然 Exchange Online 的管理员可以在 Office 365 管理中心中添加"Office 365""通讯组列表""启用邮件的安全""安全"等类型的组，但是，却无法添加"动态通讯组列表"类型的组。如果要添加"动态通讯组列表"，只能在 Exchange 管理中心中单击左侧导航栏中的"收件人"按钮，接着在右侧窗格上方单击"组"标签按钮，然后单击组列表菜单栏中的倒三角形状的"新建"按钮，并在弹出的扩展菜单中单击"动态通讯组列表"按钮来进行添加，请参见图 3-14。

然后，在弹出的"新建动态通讯组列表"对话框中键入"显示名称"和"别名"等必须填写的项目。接下来，选择成员的收件人类型，并可以单击"添加规则"按钮来添加成员身份的设定规则。例如，此处新建的动态通讯组列表的成员将包括本租户中部门名称是 Sales 的邮箱用户，如图3-18所示。

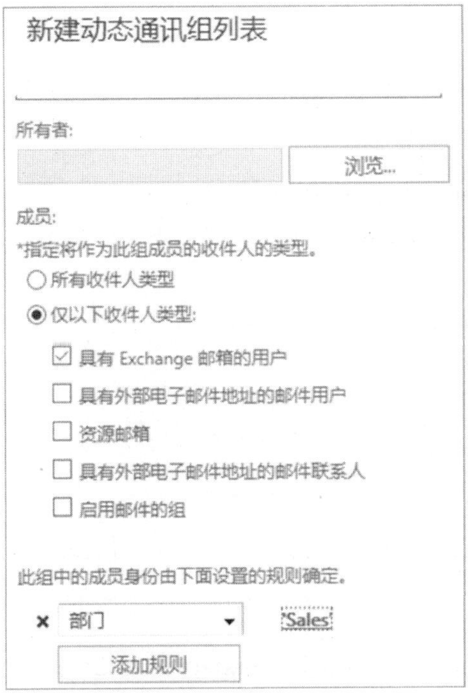

图 3-18

相应地，我们可以使用New-DynamicDistributionGroup命令来添加动态通讯组列表。参数IncludedRecipients对应"新建动态通讯组列表"对话框中"*指定将作为此组成员的收件人的类型"这个选项，其中，AllRecipients对应"所有收件人类型"，而"仅以下收件人类型"从上到下依次对应MailboxUsers、MailUsers、Resources、MailContacts、MailGroups。另外，"仅以下收件人类型"是复选框，可以使用逗号分隔的字符串进行赋值，例如：

-IncludedRecipients "MailboxUsers, MailGroups"

接下来，我们使用 PowerShell 命令来创建名称为 dynSales 的动态通讯组列表，组成员包括 Sales 部门具有 Exchange 邮箱的所有用户。命令和输出结果如下：

EXO PS > New-DynamicDistributionGroup -Name "dynSales" `
>> -IncludedRecipients "MailboxUsers" -ConditionalDepartment "Sales"

```
Name        ManagedBy
----        ---------
dynSales
```

（2）通讯组列表和启用邮件的安全组都需要设置所有者，如果我们不进行设置，组的创建者就会被设置为组的所有者，但是，动态通讯组列表的所有者并不是必须设置的，如果需要设置，则可以使用以下命令：

EXO PS > Set-DynamicDistributionGroup -Identity "dynSales" `
>> -ManagedBy "exoadmin@office365ps.partner.onmschina.cn"

在 Exchange 管理中心的组列表中单击"动态通讯组列表"的显示名称，然后单击菜单栏中的笔形"编辑"按钮；在弹出的"编辑动态通讯组列表"对话框左侧的导航栏中单击"成员身份"按钮，并在右侧的详细内容窗格中查看成员身份的设定规则。在 Exchange 管理中心中我们无法查看该组当前具体包含了哪些组成员；而使用 Get-Recipient 命令可以直接获取该组当前的组成员列表。命令和输出结果如下：

EXO PS > Get-Recipient -RecipientPreviewFilter `
>> (Get-DynamicDistributionGroup -Identity "dynSales").RecipientFilter

```
Name     RecipientType
----     -------------
user39   UserMailbox
user01   UserMailbox
user02   UserMailbox
user36   UserMailbox
user35   UserMailbox
```

如果用户 user 21 从 HR 部门调到了 Sales 部门，我们只需要更新该用户的部门信息，而不必手动维护该用户的组成员信息了。命令和输出结果如下：

EXO PS > Set-User -Identity "user21*" -Department "Sales"
EXO PS > Get-Recipient -RecipientPreviewFilter `

```
>> (Get-DynamicDistributionGroup -Identity "dynSales").RecipientFilter |
>> Where-Object{$_.Name -match "user21"}

Name    RecipientType
----    -------------
user21  UserMailbox
```

（3）与使用 Get-DistributionGroup 命令查询通讯组列表和启用邮件的安全组类似，我们可以使用 Get-DynamicDistributionGroup 命令查看租户内的动态通讯组列表。命令和输出结果如下：

EXO PS > Get-DynamicDistributionGroup

```
Name       ManagedBy
----       ---------
dynSales   exoAdmin
```

（4）另外，我们还可以使用 Remove-DynamicDistributionGroup 命令删除动态通讯组列表，必需的唯一参数仍然是该组的名称标识。命令和输出结果如下：

EXO PS > Remove-DynamicDistributionGroup -Identity "dynSales"

```
Confirm
Are you sure you want to perform this action?
Removing dynamic distribution group "dynSales".
[Y] Yes  [A] Yes to All  [N] No  [L] No to All  [?] Help (default is "Y"): Y
```

3. Office 365 组的管理

我们在使用 PowerShell 命令创建通讯组列表或启用邮件的安全组时会收到以下 Office 365 组的提示：

New! Office 365 Groups are the next generation of distribution lists.

Groups give teams shared tools for collaborating using email, files, a calendar, and more. You can start right away using the New-UnifiedGroup cmdlet.

Office 365 组有以下特点。首先，该类型的组只存在于 Exchange Online 中，Office 365 的各项服务都可以引用它；其次，该类型的组由通讯列表和共享项目组成，共享项目包括组的收件箱和日历等。用户加入该类型的组后，会被添加到该组的迪讯列表中，并可以访问该组共享项目当中的内容，甚至包括以往的邮件和共享文档等。以上这点和通讯组列表有着很大的区别，并且提升了组成员之间协作的流畅程度，因为新加入通讯组列表的成员只会接收到加入通讯组后的新邮件，而无法查看历史邮件。

（1）在组列表中，"+新建 Office 365 组"按钮或者有着同样功能的"创建组"按钮被放到菜单栏中的显著位置，如图 3-19 所示。

图 3-19

相应地，使用 PowerShell 命令 New-UnifiedGroup 即可创建 Office 365 组。需要的唯一参数只有 DisplayName。如果没有指定 Office 365 组的访问类型，会设置为默认值 Public。即本租户内的所有用户都可以访问该组的共享内容和会话，并且不经过该组所有者的审批就可以加入组。如果组的所有者要限制用户对该组的访问，可以把该组的访问类型设置为 Private，即仅该组内的成员可以访问组的共享内容和会话，并且需要经过该组所有者的审批才可以加入组。命令和输出结果如下：

EXO PS > New-UnifiedGroup -DisplayName "uniSales"

```
Name                         Alias       ServerName    AccessType
----                         -----       ----------    ----------
uniSales_501f5095-86c8... uniSales       sh2pr01mb185  Public
```

由于 Office 365 组 uniSales 的访问类型是 Public，用户 user 21 在网页版本的 Outlook 登录后，即可查询到该 uniSales 组，并可以通过单击页面右上角的"加入"按钮，来加入该组，如图 3-20 所示。

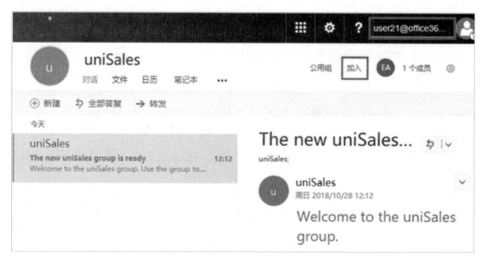

图 3-20

（2）在 Exchange 管理中心中，使用 Exchange Online 管理员的用户凭据登录，即使该 ExchangeOnline 的管理员同时也是 Office 365 组的所有者，在组列表中单击该 Office 365 组的显示名称后，组列表菜单栏中笔形的"编辑"按钮依然是以灰色显示的，即无法修改该组的各项设置。使用 Exchange Online 管理员的用户凭据登录 Office 365 的管理中心，情况也是类似的：单击左侧导航栏中的"组"按钮，在弹出的扩展菜单中再次单击"组"按钮，并在右侧的组列表中，选择该 Office 365 组的组名左侧的复选框，在页面右侧弹出的组属性对话框中，也会收到提示信息：只有全局管理员或同时具有用户管理角色的管理员可管理 Office 365 组。

使用 PowerShell 命令，以 Exchange Online 管理员的用户凭据连接 Exchange Online 后，可以对 Office 365 组的访问类型和别名等属性方便地进行修改，并进行组成员的添加、删除等操作。命令和输出结果如下：

```
EXO PS > Set-UnifiedGroup -Identity "uniHR" -AccessType Private
EXO PS > Get-UnifiedGroup -Identity "uniHR" | Select-Object DisplayName,AccessType

DisplayName AccessType
----------- ----------
uniHR       Private
```

```
EXO PS > Add-UnifiedGroupLinks -Identity "uniHR" -LinkType Members `
>> -Links user31@office365ps.partner.onmschina.cn,`
>> user32@office365ps.partner.onmschina.cn,`
>> user33@office365ps.partner.onmschina.cn
```

使用 Remove-UnifiedGroupLinks 命令可以删除 Office 365 组的组成员、所有者及该组的订阅者等信息。命令和输出结果如下：

```
EXO PS > Remove-UnifiedGroupLinks -Identity "uniHR" -LinkType Members `
>> -Links user33@office365ps.partner.onmschina.cn

Confirm
Are you sure you want to perform this action?
Removing links from unified group "uniHR" will remove the links permanently.
[Y] Yes  [A] Yes to All  [N] No  [L] No to All  [?] Help (default is "Y"): Y
```

使用 Get-UnifiedGroupLinks 命令可以查询 Office 365 组的组成员、所有者及该组的订阅者等信息。命令和输出结果如下：

```
EXO PS > Get-UnifiedGroupLinks -Identity "uniHR" -LinkType Members

Name      RecipientType
----      -------------
exoAdmin  UserMailbox
user 31   UserMailbox
user 32   UserMailbox
```

```
EXO PS > Get-UnifiedGroupLinks -Identity "uniHR" -LinkType Owner

Name        RecipientType
----        -------------
exoAdmin    UserMailbox
```

另外，以 Exchange Online 管理员的用户凭据，使用 Microsoft Azure Active Directory 模块连接 Office 365 的租户后，还可以使用以下命令查看 Office 365 组的组成员。我们要注意的是，命令 Get-MsolGroupMember 的参数 GroupObjectId，需要使用 ObjectId 类型的参数值。命令和输出结果如下：

EXO PS > $cred = Get-Credential
EXO PS > Connect-MsolService -Credential $cred -AzureEnvironment AzureChinaCloud
EXO PS > $uniGroupID = (Get-MsolGroup |
>> Where-Object {$_.DisplayName -match "uniHR"}).ObjectId
EXO PS > Get-MsolGroupMember -GroupObjectId $uniGroupID

```
GroupMemberType EmailAddress                                    DisplayName
--------------- ------------                                    -----------
User            exoAdmin@office365ps.partner.onmschina.cn       exo Admin
User            user31@office365ps.partner.onmschina.cn         user 31
User            user32@office365ps.partner.onmschina.cn         user 32
```

（3）使用 Get-UnifiedGroup 命令，我们可以仿照 Exchange 管理中心组列表的显示方式，查询并显示本租户内的 Office 365 组。命令和输出结果如下：

EXO PS > Get-UnifiedGroup |
>> Select-Object DisplayName,RecipientType,IsValid,PrimarySmtpAddress

```
DisplayName              RecipientType                   IsValid PrimarySmtpAddress
-----------              -------------                   ------- ------------------
uniHR                    MailUniversalDistributionGroup     True uniHR@office365ps.partner.onmschina.cn
uniSales                 MailUniversalDistributionGroup     True uniSales@office365ps.partner.onmschina.cn
```

（4）使用 Remove-UnifiedGroup 命令可以完成 Office 365 组的删除操作。命令和输出结果如下：

EXO PS > Remove-UnifiedGroup -Identity "uniHR"

```
Confirm
Are you sure you want to perform this action?
Removing mailbox "uniHR" will remove the Active Directory user object and mark
the mailbox and the archive (if present) in the database for removal.
[Y] Yes  [A] Yes to All  [N] No  [L] No to All  [?] Help (default is "Y"): Y
```

在我们完成删除的确认操作后，组 uniHR 在 Office 365 管理中心的组列表中就无法查看了，但是，在 Exchange 管理中心的组列表中，此类被删除的 Office 365 组却依然可以查询到，只是状态被标记为"立即删除"，如图 3-21 所示。

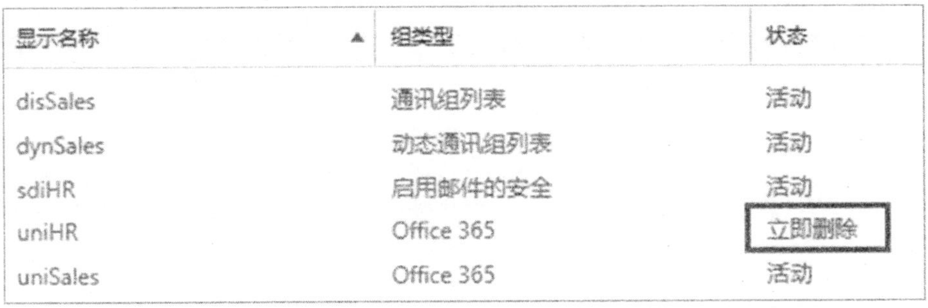

显示名称	组类型	状态
disSales	通讯组列表	活动
dynSales	动态通讯组列表	活动
sdiHR	启用邮件的安全	活动
uniHR	Office 365	立即删除
uniSales	Office 365	活动

图 3-21

与对用户邮箱进行软删除的操作类似，在 30 天的系统保留期内，Office 365 组是不会被彻底删除的，我们仍然可以查询或者恢复该 Office 365 组。命令和输出结果如下：

EXO PS > Get-UnifiedGroup -IncludeSoftDeletedGroups

```
Name                        Alias        ServerName      AccessType
----                        -----        ----------      ----------
uniHR_42d0ea25-b0e6-4b...   uniHR        sh2pr01mb0441   Private
```

EXO PS > Undo-SoftDeletedUnifiedGroup -SoftDeletedObject "uniHR"

```
Name                        Alias        ServerName      AccessType
----                        -----        ----------      ----------
uniHR_42d0ea25-b0e6-4b...   uniHR        sh2pr01mb0441   Private
```

而通讯组列表，启用邮件的安全组和动态通讯组列表被删除后，也就无法从 Office 365 管理中心或 Exchange 管理中心的组列表中进行查看了，这与 Office 365 组是有区别的。

（5）我们可以将通讯组列表升级为 Office 365 组，但是，启用邮件的安全组或者动态通讯组列表是无法升级到 Office 365 组的。在 Exchange 管理中心左侧的导航栏中单击"收件人"按钮，在右侧详细内容窗格中单击"组"标签按钮，并单击"升级通讯组列表"项下的"开始使用"按钮；然后，在弹出的"将 DL 批量升级到 Outlook 中的组"对话框中，通过选择通讯组的显示名称左侧的复选框选择一个或多个通讯组列表，并单击对话框下方的"开始升级"按钮，开始组的升级操作，如图 3-22 所示。

图 3-22

我们同样可以使用 PowerShell 命令，来进行通讯组列表的升级操作。命令如下：

EXO PS > Upgrade-DistributionGroup `
>> -DlIdentities disSales@office365ps.partner.onmschina.cn

3.2.3 如何管理资源

Exchange Online 将资源分为会议室和设备两种类型，设备资源又包括白板、录音笔、汽车等。这些资源都是以"电子邮件地址"的形式进行标识的，所以从广义上讲，也可以作为收件人对象。另外，除了利用下述方法在 Exchange 管理中心对资源进行管理，我们也可以通过 Office 365 管理中心左侧导航栏中的"资源"菜单进行类似的管理操作。

1. 会议室邮箱的管理

（1）在 Exchange 管理中心中单击左侧导航栏中的"收件人"按钮，在右侧详细内容窗格中单击上方的"资源"标签按钮；然后单击资源列表菜单栏中的"+"按钮，并在弹出的扩展菜单中单击"会议室邮箱"按钮，进行会议室邮箱资源的添加，如图 3-23 所示。

图 3-23

虽然都是使用 PowerShell 的 New-Mailbox 命令，但是创建会议室邮箱相对于创建用户邮箱来说更为简单。因为用户邮箱的创建命令包括 Name、MicrosoftOnlineServicesID 和 Password 三个必选参数，而会议室邮箱的创建只需要 Name 和 Room 两个参数，且其中的 Room 参数是开关参数，不需要进行赋值。另外，当使用 Room 参数的时候，会同时创建禁止登录的电

第 3 章 使用 PowerShell 管理 Office 365 的 Exchange Online

子邮件账号及相关联的邮箱。命令和输出结果如下：

```
EXO PS > New-Mailbox -Room -Name "CR101"

Name                                Alias                    Database                    ProhibitSendQuota     ExternalDirectoryObjectId
----                                -----                    --------                    -----------------     -------------------------
CR101                               CR101                    CHNPR01DG020-db121          49.5 GB
(53,150,2...   8cf0bdf2-319d-4949-a014-c81ffcbb...
WARNING: Failed to replicate mailbox:'CR101' to the
site:'CHNPR01.prod.partner.outlook.cn/Configuration/Sites/SH2PR01'.  The
mailbox will be available for logon in approximately 15 minutes.
```

在 Exchange 管理中心的资源列表菜单栏没有提供批量添加会议室邮箱的功能；而使用 PowerShell 的循环结构，我们可以更容易地批量创建会议室邮箱。命令和输出结果如下：

```
EXO PS > $CRs = @("CR201","CR202","CR301","CR302","CR303")
EXO PS > foreach ($CR in $CRs) {New-Mailbox -Room -Name $CR}

Name                                Alias                    Database                    ProhibitSendQuota     ExternalDirectoryObjectId
----                                -----                    --------                    -----------------     -------------------------
CR201                               CR201                    CHNPR01DG024-db034          49.5 GB
(53,150,2...   6810e0d0-26ca-4161-ad82-86950d61...
CR202                               CR202                    CHNPR01DG021-db110          49.5 GB
(53,150,2...   65f60914-09f4-4cd8-8985-07860b05...
CR301                               CR301                    CHNPR01DG021-db116          49.5 GB
(53,150,2...   d164e4bd-27b9-4d43-8eb2-539efbfd...
CR302                               CR302                    CHNPR01DG024-db034          49.5 GB
(53,150,2...   f0465673-1ec1-407e-8025-6e75c802...
CR303                               CR303                    CHNPR01DG024-db034          49.5 GB
(53,150,2...   fa3e7eba-8053-48a3-afc2-9b5b137a...
```

相应地，使用 Remove-Mailbox 命令可以删除会议室邮箱。

（2）使用 Get-Mailbox 命令，可以仿照 Exchange 管理中心资源列表的样式列出租户中的会议室邮箱。命令和输出结果如下：

```
EXO PS > Get-Mailbox | Where-Object {$_.ResourceType -match "Room"} |
>> Select-Object Name,Alias,PrimarySmtpAddress

Name  Alias PrimarySmtpAddress
----  ----- ------------------
CR101 CR101 CR101@office365ps.partner.onmschina.cn
CR201 CR201 CR201@office365ps.partner.onmschina.cn
CR202 CR202 CR202@office365ps.partner.onmschina.cn
CR301 CR301 CR301@office365ps.partner.onmschina.cn
```

```
CR302 CR302 CR302@office365ps.partner.onmschina.cn
CR303 CR303 CR303@office365ps.partner.onmschina.cn
```

（3）我们可以为会议室创建通讯组列表，并以该方式来对会议室进行分类管理。例如，可以把会议室按楼号、楼层、用途、是否配备投影仪等来进行分类。在 Exchange 管理中心的资源列表菜单中没有提供创建会议室通讯组列表的功能，需用 PowerShell 命令来创建。例如把所有位于第三层的会议室添加到会议室的通讯组列表当中。命令和输出结果如下：

EXO PS > New-DistributionGroup -Name "Cr3F" -RoomList -Members CR301,CR302

```
New! Office 365 Groups are the next generation of distribution lists.
Groups give teams shared tools for collaborating using email, files, a calendar,
and more.
You can start right away using the New-UnifiedGroup cmdlet.


Name DisplayName GroupType PrimarySmtpAddress
---- ----------- --------- ------------------
Cr3F Cr3F        Universal Cr3F@office365ps.partner.onmschina.cn
```

利用上述命令创建的会议室的通讯组列表与普通的通讯组列表还是略有区别的，例如，该会议室的通讯组列表在 Exchange 管理中心的组列表和资源列表中都无法显示，而只能通过 Get-DistributionGroup 命令查询到。我们也可以使用 Add-DistributionGroupMember 命令向会议室的通讯组列表中新添加会议室，或利用 Remove-DistributionGroupMember 命令删除会议室列表中的会议室。命令和输出结果如下：

EXO PS > Add-DistributionGroupMember -Identity "Cr3F" -Member CR303

EXO PS > Remove-DistributionGroupMember -Identity "Cr3F" -Member CR302

```
Confirm
Are you sure you want to perform this action?
Removing distribution group member "CR302" from distribution group "Cr3F".
[Y] Yes  [A] Yes to All  [N] No  [L] No to All  [?] Help (default is "Y"): Y
```

EXO PS > Get-DistributionGroupMember -Identity "Cr3F"

```
Name  RecipientType
----  -------------
CR301 UserMailbox
CR303 UserMailbox
```

类似地，我们可以使用 Remove-DistributionGroup 命令来删除会议室的通讯组列表。

2. 设备邮箱的管理

（1）企业内部能共用的设备，大到车辆，小到录音笔，都会涉及查看当前存量、状态、预约使用等操作，也可以使用创建设备邮箱的方式来进行管理。在 Exchange 管理中心中单击资源列表菜单栏中的"+"按钮，并在弹出的扩展菜单中单击"设备邮箱"按钮，来进行设备邮箱资源的添加，请参考图3-23。

与创建会议室邮箱的 PowerShell 命令类似，创建设备邮箱只需要将 Room 参数更换为 Equipment 参数即可。以下命令便为三台货车分别创建了设备邮箱。命令和输出结果如下：

```
EXO PS > $Vans = @("Van01","Van02","Van03")
EXO PS > foreach ($Van in $Vans) {New-Mailbox -Equipment -Name $Van}

Name           Alias          Database           ProhibitSendQuota        ExternalDirectoryObjectId
----           -----          --------           -----------------        -------------------------
Van01          Van01          CHNPR01DG021-db115        49.5 GB (53,150,2...
b9f939b9-6157-40c9-bf69-12368cdd...
Van02          Van02          CHNPR01DG028-db111        49.5 GB (53,150,2...
3334e00c-1622-4a19-8466-08f32b5d...
Van03          Van03          CHNPR01DG027-db096        49.5 GB (53,150,2...
82be8850-103b-4e83-a0b7-c87766b2...
```

（2）与列出租户内的会议室邮箱所使用的方法类似，我们也可以仿照 Exchange 管理中心资源列表的样式，列出租户内的设备邮箱。命令和输出结果如下：

```
EXO PS > Get-Mailbox | Where-Object {$_.ResourceType -match "Equipment"} | Select-Object Name,Alias,PrimarySmtpAddress

Name  Alias PrimarySmtpAddress
----  ----- ------------------
Van01 Van01 Van01@office365ps.partner.onmschina.cn
Van02 Van02 Van02@office365ps.partner.onmschina.cn
Van03 Van03 Van03@office365ps.partner.onmschina.cn
```

3.2.4 如何管理联系人

Exchange Online 中的联系人分为"邮件联系人"和"邮件用户"两种。

1. 邮件联系人的管理

（1）我们要添加邮件联系人，可以在 Exchange 管理中心中单击左侧导航栏中的"收件人"按钮，接着在右侧详细内容窗格的上方单击"联系人"标签按钮，然后单击联系人列表菜单栏中的"+"按钮，并在弹出的扩展菜单中单击"邮件联系人"按钮即可，如图 3-24 所示。

图 3-24

相应地，我们可以使用New-MailContact命令来添加邮件联系人。该命令中只有Name和ExternalEmailAddress这两个参数是必需的，如果Alias和DisplayName这两个参数没有被指定，会使用与Name参数相同的设置。命令和输出结果如下：

```
EXO PS > New-MailContact -Name "mc01" `
>> -ExternalEmailAddress "mc01@oe.21vianet. com"

Name                    Alias                              RecipientType
----                    -----                              -------------
mc01                    mc01                               MailContact
```

由于邮件联系人是位于企业外部的，所以每个邮件联系人都会有外部的邮件地址，并且不同的邮件联系人不能使用相同的外部邮件地址，否则就会出现如下的报错：

The proxy address "SMTP:mc01@oe.21vianet.com" is already being used by the proxy addresses or LegacyExchangeDN of "mc01". Please choose another proxy address.

批量创建邮件联系人的时候，可以使用 New-MailContact 命令，配合 PowerShell 的循环，但是，不能使用 Import-ContactList 命令。因为该命令的作用，是把从用户邮箱客户端导出的 CSV 格式的邮箱联系人文件导入该用户在 Exchange Online 中的邮箱通讯录当中的。

（2）在联系人列表中添加邮件联系人的时候，在"新建邮件联系人"对话框里是无法为邮件联系人设置国家和邮编等用户属性的。只有当邮件联系人创建完成后，单击该邮件联系人的显示名称，然后单击联系人列表菜单栏中的笔形"编辑"按钮，接着在弹出的"编辑邮件联系人"对话框中单击左侧导航栏中的"联系人信息"按钮，才可以在右侧详细内容窗格中编辑该联系人的邮政编码和街道等属性，如图 3-25 所示。

图 3-25

使用 PowerShell 为邮件联系人设置属性，会涉及两个有些相似的命令。设置上述这些与邮件设置不相关的通用属性，例如国家和邮编等，可以使用 Set-Contact 命令。命令如下：

EXO PS > Set-Contact -Identity "mc01" -PostalCode "100080" -Fax "81234567"

邮件联系人能接收的邮件格式，以及是否允许该联系人从通讯簿中隐藏等与邮件设置相关的属性，则可以使用 Set-MailContact 命令来进行修改。命令如下：

EXO PS > Set-MailContact -Identity "mc01" -HiddenFromAddressListsEnabled $True `
>> -MessageBodyFormat Text -MessageFormat Text

注意

邮件联系人能接收的邮件格式和邮件主体格式等属性是无法在图形界面中的"编辑邮件联系人"对话框中进行查看和设置的。

（3）在 Exchange 管理中心中单击左侧导航栏中的"收件人"按钮，然后单击右侧窗格上方的"联系人"标签按钮。如果在联系人列表菜单栏中没有搜索文本框，可以单击放大镜形状的"搜索"按钮使搜索文本框弹出；然后在搜索文本框中键入搜索的关键字，并单击放大镜形状的"搜索"按钮，即可查找符合条件的邮件联系人、邮件用户或来宾邮件用户，并显示这些邮件联系人或邮件用户的相关属性，如图 3-26 所示。

图 3-26

上述在联系人列表中进行搜索的操作，与使用 PowerShell 命令搜索邮件联系人并查看相关的属性有以下差异。

① 利用联系人列表菜单栏中的搜索功能可以搜索符合条件的邮件联系人、邮件用户或来宾邮件用户；而使用 Get-MailContact 命令和 Get-Contact 命令只能查询邮件联系人的信息，并且这两条命令看到的邮件联系人的属性信息是不同的。

② 虽然单击联系人列表菜单栏中的"…"按钮，在弹出的扩展菜单中单击"添加/删除列"按钮，可以在"添加/删除列"对话框中增删联系人列表的属性列，并能修改属性列的排序；但是，除了电子邮件地址和联系人类型，以及"编辑邮件联系人"对话框中"常规"属性页的"在地址列表中隐藏"属性，联系人列表无法显示更多与邮件设置相关的属性。而使用 Get-MailContact 命令，可以查看与邮件设置相关的更多属性。

③ 使用联系人列表菜单栏中的搜索功能，会用到模糊查找参数 Anr，所以搜索字符串不能以"*"开头。即使搜索字符串没有以"*"开头，也有可能因为搜索字符串的长度不足等原因导致搜索失败，或者结果匹配异常。而使用 Get-MailContact 命令和 Get-Contact 命令，并搭配支持通配符的参数 Identity，可以提高搜索的灵活度。命令和输出结果如下：

EXO PS > Get-MailContact -Identity "mc0*" |
>> Select-Object PrimarySmtpAddress, Message BodyFormat, MessageFormat

```
PrimarySmtpAddress     MessageBodyFormat  MessageFormat
------------------     -----------------  -------------
mc01@oe.21vianet.com   Text               Text
mc02@oe.21vianet.com   TextAndHtml        Mime
mc03@oe.21vianet.com   TextAndHtml        Mime
```

EXO PS > Get-Contact -Identity "mc0*" | Select-Object DisplayName,PostalCode,Fax

```
DisplayName  PostalCode  Fax
-----------  ----------  ---
mc 01        100080      81234567
mc 02        100030      82345678
mc 03        100081      83456789
```

（4）在联系人列表中单击要删除的邮件联系人的显示名称，然后在联系人列表菜单栏中单击垃圾桶形状的"删除"按钮，如图 3-27 所示；接着在弹出的"警告"对话框中单击"是"按钮，完成该邮件联系人的删除操作。

图 3-27

也可以使用 Remove-MailContact 命令来删除邮件联系人。命令和输出结果如下：

EXO PS > Remove-MailContact -Identity "mc03"

```
Confirm
Are you sure you want to perform this action?
Removing mail contact "mc03" will remove the Active Directory contact object.
[Y] Yes  [A] Yes to All  [N] No  [L] No to All  [?] Help (default is "Y"): Y
```

2. 邮件用户的管理

与邮件联系人类似，邮件用户也有外部的邮件地址，但是不同的是，邮件用户中还有 Office 365 本租户的账户，可以使用该账户凭据登录并访问本租户的资源。例如，企业的临时雇佣人员如果不想使用企业的 Exchange Online 邮件系统，而要继续使用其他邮件系统的邮箱，通过创建邮件用户的方式，该邮件用户的邮件可以被转发到其外部的邮件地址。

（1）我们要添加邮件用户，可以在 Exchange 管理中心中单击左侧导航栏中的"收件人"按钮，然后在右侧上方单击"联系人"标签按钮，接着单击联系人列表菜单栏中的"+"按钮，并在弹出的扩展菜单中单击"邮件用户"按钮即可，请参考图 3-24。

与使用 PowerShell 的命令 New-Mailbox 为用户创建邮箱类似，由于需要为邮件用户设置登录凭据，New-MailUser 命令除了需要 Name 和 MicrosoftOnlineServicesID 参数外，还需要 Password 参数，且该参数的值也必须是加密的字符串。另外，该命令还额外需要 ExternalEmailAddress 参数，而且该参数所使用的外部邮件地址不能是其他邮件用户已经注册过的。命令和输出结果如下：

```
EXO PS > New-MailUser -Name "mu02" -ExternalEmailAddress "mu02@oe.21vianet.com" `
>> -MicrosoftOnlineServicesID "mu02@office365ps.partner.onmschina.cn" `
>> -Password (ConvertTo-SecureString -String 'Office365' -AsPlainText -Force)

Name                          RecipientType
----                          -------------
mu02                          MailUser
```

（2）与利用 Set-MailContact 和 Set-Contact 命令设置邮件联系人的属性类似，我们可以使用 Set-MailUser 命令来为邮件用户设置或修改与其邮件设置相关的属性；而使用 Set-User 命令来设置或修改该邮件用户的通用属性。命令如下：

```
EXO PS > Set-MailUser -Identity "mu02" -MessageBodyFormat Text -MessageFormat Text
EXO PS > Set-User -Identity "mu02" -PostalCode "100070" -Fax "84567890"
```

要注意的是，Set-MailUser 命令虽然包括 Password 和 ResetPasswordOnNextLogon 参数，但是如果 Exchange Online 管理员使用该命令并配合上述参数为邮件用户重置密码，将会收到报错信息。命令和输出结果如下：

```
EXO PS > Set-MailUser -Identity "mu02" `
>> -Password (ConvertTo-SecureString -String 'Office365' -AsPlainText -Force) `
>> -ResetPasswordOnNextLogon $True

    Recipient "mu02" couldn't be read from domain controller "SH2PR01A001DC04.
CHNPR01A001.prod.partner.outlook.cn". This may be due to replication delays.
Switching out of Forest mode should allow this operation to complete successfully.
    + CategoryInfo          : NotSpecified: (mu02:ADObjectId) [Set-MailUser],
ManagementObjectNotFoundException
    + FullyQualifiedErrorId :
[Server=BJXPR01MB230,RequestId=2c3d568b-1a1e-4faa-a99f-f8550be5fc28,TimeStamp
=10/31/2018 6:56:11 AM]
[FailureCategory=Cmdlet-ManagementObjectNotFoundException]
C366A0F2,Microsoft.Exchange.Management.RecipientTasks.SetMailUser
    + PSComputerName        : partner.outlook.cn
```

正确的做法是，邮件用户本人使用 Set-MailUser 命令为本账户重置密码。另外，账户管理员或全局管理员可以在使用 AAD Module 连接 Office 365 的租户后，使用 Set-MsolUserPassword 命令为邮件用户重置密码。命令和输出结果如下：

```
O365 PS > Set-MsolUserPassword `
>> -UserPrincipalName "mu02@office365ps.partner.onmschina.cn" `
>> -NewPassword "Office365"

Office365
```

（3）使用 Get-MailUser 命令可以查询邮件用户及其相关属性，且这些属性多是与邮件设置相关的。另外，由于命令的输出结果中还包括来宾邮件用户（SharePoint Online 中，外部用户收到邀请并来访登录验证成功后，即成为来宾邮件用户），我们可以添加筛选条件来过滤掉这些来宾邮件用户。命令和输出结果如下：

```
EXO PS > Get-MailUser | Where-Object {$_.Name -notmatch "#EXT#"} |
>> Select-Object PrimarySmtpAddress,MessageBodyFormat,MessageFormat

PrimarySmtpAddress      MessageBodyFormat  MessageFormat
------------------      -----------------  -------------
mu01@oe.21vianet.com    Html               Mime
mu02@oe.21vianet.com    Text               Text
```

使用 Get-User 命令不仅能查询邮件用户及其通用属性，也能查询除邮件联系人外的其他类型的用户及其通用属性，其他类型的用户包括会议室邮箱、设备邮箱和来宾邮件用户等。命令和输出结果如下：

```
EXO PS > Get-User |
>> Where-Object{($_.RecipientType -eq "MailUser") -and ($_.Name -notmatch "#EXT#")} |
>> Select-Object DisplayName,PostalCode,Fax

DisplayName  PostalCode  Fax
-----------  ----------  ---
mu 01        100060      85678901
mu 02        100070      84567890
```

（4）使用 Remove-MailUser 命令可以删除邮件用户。命令和输出结果如下：

```
EXO PS > Remove-MailUser -Identity "mu02"

Confirm
Are you sure you want to perform this action?
Removing the mail enabled user "mu02" will delete the mail enabled user and
the associated Windows Live ID "mu02@office365ps.partner.onmschina.cn".
[Y] Yes  [A] Yes to All  [N] No  [L] No to All  [?] Help (default is "Y"): Y
```

3.2.5 如何管理共享邮箱

企业中对外提供服务或对内提供支持的部门，例如客户服务、信息系统、人力资源等，通常会有设立部门邮箱的需求，这样本部门的员工便可以登录本部门的邮箱，共同处理发送到部门邮件地址的邮件。在本地部署的 Exchange 服务器，例如 Exchange 2016 上，通常的做法是添加用户邮箱，并为要访问该邮箱的员工，在其邮件客户端软件上配置该邮箱的用户凭据。而在 Exchange Online 上，如果仿照之前的做法添加用户邮箱，就意味着要额外消耗许可证。为了不占用许可证的资源，我们可以使用共享邮箱来实现该需求。由于共享邮箱没有用户凭据，无法使用登录邮箱的方法来访问该邮箱，只有给要访问邮箱的用户账户授予该共享邮箱的相应权限，该用户才能访问共享邮箱。

1. 共享邮箱的添加和删除

（1）在 Exchange 管理中心中单击左侧导航栏中的"收件人"按钮，然后在右侧详细内容窗格中单击上方的"共享"标签按钮，并在共享邮箱菜单栏中单击"+"按钮来新建共享邮箱，如图 3-28 所示。

图 3-28

添加共享邮箱也是使用 PowerShell 的 New-Mailbox 命令，但是需要搭配使用 Shared 参数，而无须使用 Password 密码参数。命令和输出结果如下：

EXO PS > New-Mailbox -Name "shareSales" -Shared

```
Name                              Alias                 Database
ProhibitSendQuota    ExternalDirectoryObjectId
----                              -----                 --------
----------------     -------
shareSales                        shareSales            CHNPR01DG025-db039              49.5 GB
(53,150,2... ee51a03d-d416-48c6-a199-6cb6a068...
   WARNING: Failed to replicate mailbox:'shareSales' to the
```

```
site:'CHNPR01.prod.partner.outlook.cn/Configuration/Sites/SHAPR01'. The
mailbox will be available for logon in approximately 15 minutes.
```

（2）如果在共享邮箱列表的菜单栏中没有搜索文本框，单击放大镜形状的"搜索"按钮可以使该文本框弹出。在搜索文本框中键入关键字，并单击该搜索文本框中放大镜形状的"搜索邮箱"按钮，即可查看符合搜索条件的共享邮箱，如图3-29所示。

图 3-29

使用共享邮箱列表菜单栏中的"搜索"功能与"高级搜索"功能，将会遇到与用户邮箱列表菜单栏中的搜索功能同样的问题，即搜索字符串无法以"*"开头、无法精确匹配字符串等；而使用 Get-Mailbox 命令查询租户中的共享邮箱，通过搭配使用 Identity 参数，可以解决上述共享邮箱搜索方面的问题。由于 Get-Mailbox 命令除了可以查询共享邮箱，还可以查询其他类型的邮箱，例如用户邮箱、会议室邮箱等，我们还需要额外添加条件来对邮箱的类型进行过滤。命令和输出结果如下：

EXO PS > Get-Mailbox -Identity "*Sales" |
>> Where-Object {($_.IsShared -eq "True") -and ($_.Name -notmatch "DiscoverySearch")}

```
Name                               Alias                      Database
ProhibitSendQuota    ExternalDirectoryObjectId
----                               -----                      --------
----------------     -------------------------
    shareSales                     shareSales                 CHNPR01DG025-db039          49.5 GB
(53,150,2...  ee51a03d-d416-48c6-a199-6cb6a068...
```

在 Exchange 管理中心的共享邮箱列表中单击需要删除的共享邮箱的显示名称，然后单击菜单栏中垃圾桶形状的"删除"按钮即可删除该共享邮箱。类似地，使用 Remove-Mailbox 命令可以删除共享邮箱。命令和输出结果如下：

EXO PS > Remove-Mailbox -Identity "shareHR"

```
Confirm
Are you sure you want to perform this action?
Removing the mailbox "shareHR" will mark the mailbox and the archive, if
present, for deletion. The associated Windows Live ID
"shareHR@office365ps.partner.onmschina.cn" will also be deleted and will not
be available for any other Windows Live service.
[Y] Yes  [A] Yes to All  [N] No  [L] No to All  [?] Help (default is "Y"): Y
```

第 3 章 使用 PowerShell 管理 Office 365 的 Exchange Online

虽然被删除的共享邮箱会立即从 Exchange 管理中心的共享邮箱列表中消失，但是在 Office 365 管理中心左侧的导航栏中单击"用户"按钮，并在弹出的扩展菜单中单击"已删除的用户"按钮，然后可以在右侧的已删除用户列表中查看该被删除的共享邮箱账户。我们也可以使用下面的 PowerShell 命令来查询已经被删除的共享邮箱，而且在 30 天的系统保留期内，我们可以选择恢复该共享邮箱。命令和输出结果如下：

```
EXO PS > Get-Mailbox -SoftDeletedMailbox | Where-Object {$_.IsShared -eq "True"}

Name                          Alias                  Database
ProhibitSendQuota    ExternalDirectoryObjectId
----                          -----                  --------
-----------------    -------------------------
shareHR                       shareHR                CHNPR01DG035-db083              49.5 GB
(53,150,2... a91842b5-3885-46ce-b666-888869c5...
```

2. 共享邮箱的权限管理

在 Exchange 管理中心的共享邮箱列表中单击要配置或修改权限的共享邮箱的显示名称以选择该邮箱，接着单击菜单栏中的笔形"编辑"按钮，然后在弹出的编辑共享邮箱对话框中单击左侧导航栏中的"邮箱委托"按钮，并在右侧详细内容窗格中，配置或修改"完全访问"或"发送方式"的权限，如图 3-30 所示。

图 3-30

（1）由于用户无法用登录邮箱的方法访问该邮箱，需要用 Add-Mailbox Permission 命令为要访问该共享邮箱的用户账户授予权限。以下命令将完全访问权限授予用户 user 31，命令成功执行后，该用户不仅可以访问该共享邮箱，阅读邮箱中的邮件，创建联系人等，甚至可以进行删除邮件的操作。另外，在编辑共享邮箱对话框中的"邮箱委托"属性项下，我们无法为用户授予除完全访问权限外的其他权限，如果要为用户授予 ReadPermission 或 ChangePermission 等其他类型的权限，我们只能通过使用 Add-MailboxPermission 这个 PowerShell 命令。命令和输出结果如下：

EXO PS > Add-MailboxPermission -Identity "shareSales" -User "user31" `
\>\> -AccessRights FullAccess

Identity	User	AccessRights	IsInherited	Deny
shareSales	CHNPR01A001\user2...	{FullAccess}	False	False

要想在 Outlook 网页版中访问该共享邮箱，可以在浏览器地址栏中直接键入以下 URL：https://partner.outlook.cn/owa/。

在键入用户凭据成功登录后，单击页面右上角的用户头像按钮，并在弹出的"我的账户"菜单中单击"打开其他邮箱..."按钮来访问共享邮箱，如图 3-31 所示。

图 3-31

我们可以使用 Get-MailboxPermission 命令来查询有哪些用户在该共享邮箱上被授予了何种级别的访问权限，而在编辑共享邮箱对话框中的"邮箱委托"属性项下，我们仅能查看该共享邮箱上被授予了完全访问权限的用户账户。另外，该查询命令的输出结果中，还包括系统账户所拥有的访问权限。命令和输出结果如下：

EXO PS > Get-MailboxPermission -Identity "shareSales" |

```
>> Select-Object User,AccessRights
```

```
User                                       AccessRights
----                                       ------------
NT AUTHORITY\SELF                          {FullAccess, ReadPermission}
NT AUTHORITY\SELF                                   {FullAccess, ExternalAccount,
ReadPermission}
user21@office365ps.partner.onmschina.cn {FullAccess}
user31@office365ps.partner.onmschina.cn {FullAccess}
user34@office365ps.partner.onmschina.cn {FullAccess}
user33@office365ps.partner.onmschina.cn {FullAccess}
CHNPR01\Administrator                      {FullAccess}
CHNPR01\Domain Admins                      {FullAccess}
CHNPR01\Enterprise Admins                  {FullAccess}
CHNPR01\Organization Management            {FullAccess}
NT AUTHORITY\SYSTEM                        {FullAccess}
NT AUTHORITY\NETWORK SERVICE               {ReadPermission}
CHNMGT03\JitUsers                          {ReadPermission}
CHNPR01\Administrator                      {FullAccess, DeleteItem, ReadPermission,
ChangePermission, ChangeOwner}
CHNPR01\Domain Admins                      {FullAccess, DeleteItem, ReadPermission,
ChangePermission, ChangeOwner}
CHNPR01\Enterprise Admins                  {FullAccess, DeleteItem, ReadPermission,
ChangePermission, ChangeOwner}
CHNPR01\Organization Management            {FullAccess, DeleteItem, ReadPermission,
ChangePermission, ChangeOwner}
CHNPR01\Public Folder Management           {ReadPermission}
CHNPR01\Exchange Servers                   {FullAccess, ReadPermission}
CHNPR01\Exchange Trusted Subsystem         {FullAccess, DeleteItem, ReadPermission,
ChangePermission, ChangeOwner}
CHNPR01\Managed Availability Servers       {ReadPermission}
```

我们可以使用 Remove-MailboxPermission 命令从指定的共享邮箱上移除特定用户的访问权限。命令和输出结果如下：

```
EXO PS > Remove-MailboxPermission -Identity "shareSales" `
>> -User "user21@office365ps.partner.onmschina.cn" -AccessRights FullAccess
```

```
Confirm
Are you sure you want to perform this action?
Removing mailbox permission "shareSales" for user
"user21@office365ps.partner.onmschina.cn" with access rights "'FullAccess'".
[Y] Yes  [A] Yes to All  [N] No  [L] No to All  [?] Help (default is "Y"): Y
```

（2）即使已经为用户授予了共享邮箱的完全访问权限，如果该用户从 Outlook 网页版或邮件客户端软件对邮箱中的邮件进行回复操作，该用户也会收到警告：您没有从此邮箱发送邮件的权限。我们需要使用 Set-Mailbox 命令，而且也只能通过该命令，为该用户另外授予代

表共享邮箱发送邮件的权限,来解决上述问题。命令如下:

```
EXO PS > Set-Mailbox -Identity "shareSales" `
>> -GrantSendOnBehalfTo "user31@office365ps.partner.onmschina.cn",`
>> "user33@office365ps.partner.onmschina.cn"
```

上述命令成功执行后,用户 user 31 和 user 33 即可代表该共享邮箱回复发送到该共享邮箱的邮件了。

注意

参数 GrantSendOnBehalfTo 的属性值必须是本租户中的邮箱用户、邮件用户或有邮箱属性的安全组,而具体的属性值只要是可以唯一代表上述对象的标识符即可。

被回复的邮件会显示:

user 31 <user31@office365ps.partner.onmschina.cn>; on behalf of shareSales <shareSales@office365ps.partner.onmschina.cn> 。

我们可以使用 Get-Mailbox 命令查询特定的共享邮箱上有哪些用户账户被授予了代表共享邮箱发送邮件的权限,而且可以通过 Set-Mailbox 命令来修改或清空被授予该项权限的用户列表,修改用户列表的方式是使用新的用户列表替代旧列表。虽然共享邮箱的 GrantSendOnBehalfTo 属性的属性值在命令执行完毕后随即发生改变,但是用户账户权限的生效或撤销仍然需要一些时间。命令和输出结果如下:

```
EXO PS > Get-Mailbox -Identity "shareSales" | Select-Object Name,GrantSendOnBehalfTo

Name         GrantSendOnBehalfTo
----         -------------------
shareSales   {user31, user33}
```

```
EXO PS > Set-Mailbox -Identity "shareSales" `
>> -GrantSendOnBehalfTo "user33@office365ps.partner.onmschina.cn",`
>> "user34@office365ps.partner.onmschina.cn"
EXO PS > Get-Mailbox -Identity "shareSales" |
>> Select-Object Name,GrantSendOnBehalfTo

Name         GrantSendOnBehalfTo
----         -------------------
shareSales   {user33, user34}
```

```
EXO PS > Set-Mailbox -Identity "shareSales" -GrantSendOnBehalfTo $null
EXO PS > Get-Mailbox -Identity "shareSales" |
>> Select-Object Name,GrantSendOnBehalfTo

Name         GrantSendOnBehalfTo
----         -------------------
shareSales   {}
```

（3）如果用户在回复共享邮箱中的邮件时，需要被回复邮件的发件人显示为来自共享邮箱 shareSales<shareSales@office365ps.partner.onmschina.cn>，而非来自 user 31 on behalf of shareSales，仅凭为该用户添加代表共享邮箱发送邮件的权限是无法实现的。我们需要在编辑共享邮箱对话框中的"邮箱委托"项下将该用户添加到"发送方式"的权限列表中。使用 Add-RecipientPermission 命令也可以为用户账户添加 SendAs 权限。命令和输出结果如下：

```
EXO PS > Add-RecipientPermission -Identity "shareSales" -AccessRights SendAs `
>> -Trustee "user33@office365ps.partner.onmschina.cn"

Confirm
Are you sure you want to perform this action?
Adding recipient permission 'SendAs' for user or group
'user33@office365ps.partner.onmschina.cn' on recipient 'shareSales'.
[Y] Yes  [A] Yes to All  [N] No  [L] No to All  [?] Help (default is "Y"): Y

Identity    Trustee  AccessControlType AccessRights Inherited
--------    -------  ----------------- ------------ ---------
shareSales  user33   Allow             {SendAs}     False
```

使用 Get-RecipientPermission 命令可以查询在特定的共享邮箱上，有哪些用户被赋予了 SendAs 的权限。命令和输出结果如下：

```
EXO PS > Get-RecipientPermission -Identity "shareSales"

Identity    Trustee                                    AccessControlType AccessRights Inherited
--------    -------                                    ----------------- ------------ ---------
shareSales  NT AUTHORITY\SELF                          Allow             {SendAs}     False
shareSales  user33@office365ps.partner.onmschina.cn    Allow             {SendAs}     False
```

使用 Remove-RecipientPermission 命令可以移除指定用户在该共享邮箱上的 SendAs 权限。命令和输出结果如下：

```
EXO PS > Remove-RecipientPermission -Identity "shareSales" -AccessRights SendAs `
>> -Trustee "user33@office365ps.partner.onmschina.cn"

Confirm
Are you sure you want to perform this action?
Removing recipient permission 'SendAs' for user or group
'user33@office365ps.partner.onmschina.cn' from recipient 'shareSales'.
[Y] Yes  [A] Yes to All  [N] No  [L] No to All  [?] Help (default is "Y"): Y
```

3. 共享邮箱其他属性的设置

共享邮箱其他属性的设置与用户邮箱的属性设置是类似的，具体的设置方法请参考本章前面的"3.2.1 邮箱的管理"中的"3. 修改用户邮箱的属性"部分的内容。

3.3 如何使用 PowerShell 管理 Exchange Online 的权限

Exchange Online 的权限管理包括对管理员角色的管理、对用户角色的管理及对 OWA 策略的管理。我们通过将用户作为成员加入不同的管理员角色组来为特定用户分配相应的管理权限，而通过修改默认的用户角色策略和 Outlook Web App 策略，或通过新建相应的策略，来为用户在 Exchange Online 中所能进行操作，以及在 OWA 中可以使用的功能进行授权。

3.3.1 如何管理管理员角色

1. 管理员角色组的查看

从 Exchange 2010 版本开始，Exchange Server 就引入了 RBAC 的概念，即基于角色的访问控制。在该权限模型中，管理员角色组是一种特殊的通用安全组。

（1）在 Exchange 管理中心中单击左侧导航栏中的"权限"按钮，然后单击右侧详细内容窗格上方的"管理员角色"标签按钮，即可查看本租户内所有的管理员角色组；并且可以单击特定角色组的名称以选择该角色组，然后单击菜单栏中的笔形"编辑"按钮，并在弹出的编辑角色组属性对话框中查看该角色组的详细信息，如图 3-32 所示。

图 3-32

相应地，我们可以使用不带参数的 Get-RoleGroup 命令查询租户内包含的管理员角色组。命令和输出结果如下：

EXO PS > Get-RoleGroup | Select-Object Name,Roles

```
Name                                             Roles
----                                             -----
```

```
    Organization Management                              {Address Lists,
ApplicationImpersonation, ArchiveApplication, Audit Logs...}
    Recipient Management                                 {Distribution Groups, Mail
Recipient Creation, Mail Recipients, Message Trac...
    View-Only Organization Management                    {View-Only Configuration,
View-Only Recipients}
    UM Management                                        {UM Mailboxes, UM Prompts,
Unified Messaging}
    Help Desk                                            {Reset Password, User Options,
View-Only Recipients}
    Records Management                                   {Audit Logs, Journaling, Message
Tracking, Retention Management...}
    Discovery Management                                 {Legal Hold, Mailbox Search}
    Hygiene Management                                   {Transport Hygiene, View-Only
Configuration, View-Only Recipients}
    Compliance Management                                {Audit Logs, Compliance Admin,
Data Loss Prevention, Information Rights Mana...
    O365 Support View Only                               {O365SupportViewConfig}
    Security Reader                                      {Security Reader}
    Security Administrator                               {Security Admin}
    HelpdeskAdmins_-1748086685                           {}
    TenantAdmins_192512474                               {}
    ComplianceAdmins_729475203                           {}
    ExchangeServiceAdmins_-1763236442                    {}
    SecurityAdmins_-1324667764                           {}
    RIM-MailboxAdmins10e9251bfd0d427290dd4a77a073d8e5
{ApplicationImpersonation}
```

（2）管理员角色组是把相对琐碎的多种角色，按照类别集合在一起，以简化权限的分配操作。角色组包含角色、与角色匹配的写入作用域、该角色组包括的成员等属性。我们同样可以利用 Get-RoleGroup 命令，查询特定管理员角色组中的所有角色。由于该命令输出结果中 Roles 属性的部分属性值含有较长的字符串，这些超过标准列宽的属性值字符串将会被截断，所以我们通过添加 ExpandProperty 参数来保留全部的属性值字符串。命令和输出结果如下：

```
EXO PS > Get-RoleGroup -Identity "Recipient Management" |
>> Select-Object -ExpandProperty Roles

    Migration
    Message Tracking
    Mail Recipients
    Move Mailboxes
    Mail Recipient Creation
    Distribution Groups
    Reset Password
    Team Mailboxes
```

使用 Get-RoleGroupMember 命令来查询特定的管理员角色组中包括的成员，而针对 Recipient Management 这个角色组，以下命令执行后没有返回任何的结果，即该角色组不包含任何的成员。命令如下：

EXO PS > Get-RoleGroupMember -Identity "Recipient Management"

管理员角色组中的每个特定角色可以设置不同于其他角色的写入作用域，并以 RoleAssignment 为单位进行管理操作，且每个 RoleAssignment 都有唯一的名称，命名方式为：管理员角色-管理员角色组。我们可以使用 Get-ManagementRoleAssignment 命令来查询角色组中的 RoleAssignment，并进一步查询该特定 RoleAssignment 对应角色的写入作用域。命令和输出结果如下：

EXO PS > Get-ManagementRoleAssignment -RoleAssignee "Recipient Management" |
>> Select-Object Name,Role,RoleAssigneeName,RoleAssigneeType

```
Name                                         Role                      RoleAssigneeName    RoleAssigneeType
----                                         ----                      ----------------    ----------------
Migration-Recipient Management               Migration                 Recipient Management   RoleGroup
Message Tracking-Recipient Management        Message Tracking          Recipient Management   RoleGroup
Mail Recipients-Recipient Management         Mail Recipients           Recipient Management   RoleGroup
Move Mailboxes-Recipient Management          Move Mailboxes            Recipient Management   RoleGroup
Mail Recipient Creation-Recipient Management Mail Recipient Creation   Recipient Management   RoleGroup
Distribution Groups-Recipient Management     Distribution Groups       Recipient Management   RoleGroup
Reset Password-Recipient Management          Reset Password            Recipient Management   RoleGroup
Team Mailboxes-Recipient Management          Team Mailboxes            Recipient Management   RoleGroup
Recipient Policies-Recipient Management      Recipient Policies        Recipient Management   RoleGroup
```

EXO PS > Get-ManagementRoleAssignment `
>> -Identity "Mail Recipients-Recipient Management" |
>> Select-Object RecipientWriteScope,CustomRecipientWriteScope

```
RecipientWriteScope  CustomRecipientWriteScope
-------------------  -------------------------
Organization
```

（3）在 RBAC 权限模型中，权限的控制粒度细化到了 PowerShell 命令，甚至是 PowerShell

命令的参数。我们可以使用 Get-ManagementRoleEntry 命令查询特定管理员角色中包含的 Exchange Online 命令，而以下这些信息我们是无法在图形界面的编辑角色组属性对话框中查看到的。命令和输出结果如下：

```
EXO PS > Get-ManagementRoleEntry -Identity "Mail Recipients\*"

Name                              Role                        Parameters
----                              ----                        ----------
Set-MailboxMessageConfigura...    Mail Recipients             {AfterMoveOrDeleteBehavior,
AlwaysShowBcc, AlwaysShowFrom, AutoAddSig...
Reset-TxpBlockStatus              Mail Recipients             {Confirm, ErrorAction,
ErrorVariable, Identity...}
New-UnifiedGroup                  Mail Recipients             {AccessType, Alias,
AlwaysSubscribeMembersToCalendarEvents, AutoSubsc...
Set-UnifiedGroup                                              Mail Recipients
{AcceptMessagesOnlyFromSendersOrMembers, AccessType, Alias, AlwaysSub...
Set-Place                         Mail Recipients             {AudioDeviceName,
Building, Capacity, City...}
Set-OwaMailboxPolicy                                          Mail Recipients
{ActionForUnknownFileAndMIMETypes, ActiveSyncIntegrationEnabled, AllA...
Set-TxpUserSettings               Mail Recipients             {Confirm, ErrorAction,
ErrorVariable, Identity...}
Set-Mailbox                       Mail Recipients             {AcceptMessagesOnlyFrom,
AcceptMessagesOnlyFromDLMembers, AcceptMessa...
……
New-SweepRule                     Mail Recipients             {Confirm,
DestinationFolder, Enabled, ErrorAction...}
New-OwaMailboxPolicy              Mail Recipients             {Confirm, ErrorAction,
ErrorVariable, Name...}
New-Mailbox                       Mail Recipients             {EnableRoomMailboxAccount}
New-InboxRule                     Mail Recipients             {AlwaysDeleteOutlookRulesBlob,
ApplyCategory, ApplySystemCategory, Bo...
Import-RecipientDataProperty      Mail Recipients             {Confirm,
ErrorAction, ErrorVariable, FileData...}
Import-ContactList                Mail Recipients             {Confirm, CSV, CSVData,
CSVStream...}
……
Get-MailboxCalendarConfigur...    Mail Recipients             {ErrorAction,
ErrorVariable, Identity, OutBuffer...}
Get-MailboxAutoReplyConfigu...    Mail Recipients             {ErrorAction,
ErrorVariable, Identity, OutBuffer...}
Get-Mailbox                       Mail Recipients             {Anr, Archive, Async,
ErrorAction...}
Get-MailUser                      Mail Recipients             {Anr, ErrorAction,
ErrorVariable, Filter...}
Get-MailContact                   Mail Recipients             {Anr, ErrorAction,
```

```
ErrorVariable, Filter...}
   Get-LogonStatistics          Mail Recipients         {ErrorAction,
ErrorVariable, Identity, OutBuffer...}
```

2. 管理员角色组的创建

在 Exchange 管理中心中单击左侧导航栏中的"权限"按钮，接着单击右侧窗格上方的"管理员角色"标签按钮，并单击角色组列表菜单栏中的"+"按钮，即可新建管理员角色组，请参考图 3-32。然而，利用"新建角色组"对话框新建角色组时，我们仅能为角色组选择并添加系统默认定义的角色。

将普通用户加入上述包含多种角色的管理员角色组，容易导致的问题是，会使该用户被授予该角色组中部分角色所具有的不适合该用户的权限，或者被允许运行角色中部分该用户不能拥有执行权限的命令，从而违背了系统管理员应该遵守的最小特权原则，同时也与基于角色访问控制的设计初衷相违背。例如，我们要让用户 user 01 协助管理会议室邮箱，而又不能管理除会议室邮箱外其他类型的收件人，该需求是无法在 Exchange 管理中心的角色组列表中通过添加新角色或修改现有角色来实现的。

为了实现上述需求，我们通过执行下面的步骤用 PowerShell 命令给该用户授权。

我们先进行以下测试，使用 user 01 的用户凭据创建一个与 Exchange Online 的连接，并尝试执行 New-Mailbox 命令，以确认该用户目前是没有权限使用 New-Mailbox 命令的。命令和输出结果如下：

USER01 PS > New-Mailbox -Room -Name "CR506"

```
New-Mailbox : The term 'New-Mailbox' is not recognized as the name of a cmdlet,
function, script file, or operable program. Check the spelling of the name, or
if a path was included, verify that the path is correct and try again.
At line:1 char:1
+ New-Mailbox -Room -Name "CR506"
+ ~~~~~~~~~~~
    + CategoryInfo          : ObjectNotFound: (New-Mailbox:String) [],
CommandNotFoundException
    + FullyQualifiedErrorId : CommandNotFoundException
```

（1）我们首先新建一个自定义的管理员角色，并且该自定义角色只能使用三个与 Mailbox 有关的命令。命令和输出结果如下：

EXO PS > New-ManagementRole -Name "OnlyNewCr" `
>> -EnabledCmdlets @("New-Mailbox","Get-Mailbox","Remove-Mailbox") `
>> -Parent "Mail Recipient Creation"

```
Name      RoleType
----      --------
OnlyNewCr MailRecipientCreation
```

EXO PS > Get-ManagementRoleEntry -Identity "OnlyNewCr*"

```
Name                          Role                    Parameters
----                          ----                    ----------
Get-Mailbox                   OnlyNewCr               {Anr, Archive, ErrorAction,
ErrorVariable...}
New-Mailbox                   OnlyNewCr               {ActiveSyncMailboxPolicy,
Alias, Archive, Confirm...}
Remove-Mailbox                OnlyNewCr               {Confirm, ErrorAction,
ErrorVariable, Force...}
```

（2）然后，我们使用 New-ManagementScope 命令创建一个自定义的管理范围，其中限定了收件人的类型仅可以是 RoomMailbox，即会议室邮箱，而不是其他类型的收件人。命令和输出结果如下：

EXO PS > New-ManagementScope -Name "CrCreator" `
\>> -RecipientRestrictionFilter {RecipientTypeDetails -eq "RoomMailbox"}

```
Name          ScopeRestrictionType   Exclusive   RecipientRoot   RecipientFilter
ServerFilter
----          --------------------   ---------   -------------   ---------------
------------
CrCreator     RecipientScope         False                       RecipientTypeDetails -eq
'RoomMailbox'
```

（3）最后，创建一个新的管理员角色组，添加之前创建的 OnlyNewCr 角色，设置 CrCreator 为自定义的写入作用域，并设置这个角色组的成员。命令和输出结果如下：

EXO PS > New-RoleGroup -Name "GrpOnlyNewCr" -Roles "OnlyNewCr" `
\>> -CustomRecipientWriteScope "CrCreator" -Members "user01"

```
Name                                                              AssignedRoles
RoleAssignments                                                   ManagedBy
----                                                              -------------
---------------                                                   ---------
GrpOnlyNewCr                                                      {OnlyNewCr}
{office365ps.partner.onmschina.cn\OnlyNewCr-GrpOnlyNewCr}         {Organization
Management, exoAdmin}
```

虽然上述命令执行后很快返回了执行成功的结果，但是如果此时以 user 01 的用户凭据连接 Exchange Online，该用户依然是没有使用 New-Mailbox 命令的权限的。该用户需要等待 15 分钟或更长的时间，新授予给该用户的权限才会生效。权限生效后，该用户可以重新打开一个 PowerShell 的控制台窗口，新建与 Exchange Online 的连接，并执行以下命令。命令和输出结果如下：

USER01 PS > New-Mailbox -Room -Name "CR506"

```
Name          Alias          Database                 ProhibitSendQuota
ExternalDirectoryObjectId
```

```
    ----            -----           --------                        ------------------
    ------------------------
    CR506           CR506           CHNPR01DG028-db115              49.5 GB (53,150,2...
b4574cdf-399a-4f6c-8767-b8f939c7...
    WARNING: Failed to replicate mailbox:'CR506' to the
    site:'CHNPR01.prod.partner.outlook.cn/Configuration/Sites/SH2PR01'. The
    mailbox will be available for logon in approximately 15 minutes.
```

虽然该用户可以成功创建会议室邮箱，但是如果其尝试创建其他类型的收件人，例如尝试创建用户邮箱，依然会收到"写入范围非法"的错误警告。命令和输出结果如下：

USER01 PS > New-Mailbox -Name "user40" `
>> -MicrosoftOnlineServicesID "user40@ office365ps.partner.onmschina.cn" `
>> -Password (ConvertTo-SecureString -String 'Office365' -AsPlainText -Force) `
>> -FirstName "user" -LastName "40"

```
    WARNING: After you create a new mailbox, you must go to the Office 365 Admin
Center and assign the mailbox a license, or it will be disabled after the grace
period.
    The following error occurred during validation in agent 'Windows LiveId Agent':
'Unable to perform the save operation.
'CHNPR01A001.prod.partner.outlook.cn/Microsoft Exchange Hosted
Organizations/office365ps.partner.onmschina.cn/user40' is not within a valid
server write scope.'
        + CategoryInfo                  : NotSpecified: (:) [New-Mailbox],
ProvisioningValidationException
        + FullyQualifiedErrorId :
[Server=BJBPR01MB051,RequestId=63707d2e-0ed1-4e8c-910d-e93fbe92b83b,TimeStamp
=11/3/2018 12:00:04 PM] [FailureCategory=Cmdlet-ProvisioningValidationException]
18BC0109,Microsoft.Exchange.Management.RecipientTasks.NewMailbox
        + PSComputerName          : partner.outlook.cn
```

3. 管理员角色组属性的修改

（1）我们无法从 Exchange Online 的默认管理员角色中删除该角色包含的 Exchange Online 命令，而只能从自定义的管理员角色中删除其包含的 Exchange Online 命令，来实现控制用户权限的目标。命令和输出结果如下：

EXO PS > Remove-ManagementRoleEntry -Identity "OnlyNewCr\Remove-Mailbox"

```
    Confirm
    Are you sure you want to perform this action?
    Removing the "(Microsoft.Exchange.Management.PowerShell.E2010)
Remove-Mailbox
    -Confirm -ErrorAction -ErrorVariable -Force -Identity -IgnoreLegalHold
    -Migration -OutBuffer -OutVariable -PermanentlyDelete -PublicFolder
    -WarningAction -WarningVariable -WhatIf" management role entry on the
```

```
"OnlyNewCr" management role.
[Y] Yes  [A] Yes to All  [N] No  [L] No to All  [?] Help (default is "Y"): Y
```

EXO PS > Get-ManagementRoleEntry -Identity "OnlyNewCr*"

```
Name                            Role                       Parameters
----                            ----                       ----------
Get-Mailbox                     OnlyNewCr                  {Anr, Archive, ErrorAction,
ErrorVariable...}
New-Mailbox                     OnlyNewCr                  {ActiveSyncMailboxPolicy,
Alias, Archive, Confirm...}
```

（2）我们可以使用 New-ManagementRoleAssignment 命令给管理员角色组添加角色。下述命令首先创建自定义的角色 GetPF，该角色只包含 Get-PublicFolder 命令。新角色创建完成后，使用默认的写入作用域将 GetPF 角色添加到管理员角色组 GrpOnlyNewCr。命令和输出结果如下：

```
EXO PS > New-ManagementRole -Name "GetPF" `
>> -EnabledCmdlets @("Get-PublicFolder") -Parent "Public Folders"

Name  RoleType
----  --------
GetPF PublicFolders
```

```
EXO PS > New-ManagementRoleAssignment -SecurityGroup "GrpOnlyNewCr" `
>> -Role "GetPF"

Name                      Role              RoleAssigneeName    RoleAssigneeType
AssignmentMethod  EffectiveUserName
----                      ----              ----------------    ----------------
----------------  -----------------
GetPF-GrpOnlyNewCr        GetPF             GrpOnlyNewCr        RoleGroup
Direct
```

```
EXO PS > Get-ManagementRoleAssignment -RoleAssignee "GrpOnlyNewCr"

Name                      Role              RoleAssigneeName    RoleAssigneeType
AssignmentMethod  EffectiveUserName
----                      ----              ----------------    ----------------
----------------  -----------------
OnlyNewCr-GrpOnlyNewCr    OnlyNewCr         GrpOnlyNewCr        RoleGroup
Direct            All Group Members
GetPF-GrpOnlyNewCr        GetPF             GrpOnlyNewCr        RoleGroup
Direct            All Group Members
```

相应地，我们可以使用 Remove-ManagementRoleAssignment 命令和 Identity 参数来将特定的管理员角色从角色组中移除，以达到缩减角色组权限范围的目的。另外，我们还可以使用 Remove-ManagementRole 命令和 Identity 参数从本租户中彻底删除以前创建的自定义角色。

（3）我们可以使用替换写入作用域的方法，来达到修改写入作用域的目的。例如，我们将创建一个新的自定义写入作用域 EqpCreator，并使用该自定义写入作用域替换以前定义的作用域 CrCreator。命令和输出结果如下：

EXO PS > New-ManagementScope -Name "EqpCreator" `
>> -RecipientRestrictionFilter {RecipientTypeDetails -eq "EquipmentMailbox"}

```
Name               ScopeRestrictionType  Exclusive  RecipientRoot   RecipientFilter
ServerFilter
----               --------------------  ---------  -------------   ---------------
------------
EqpCreator         RecipientScope        False                      RecipientTypeDetails
-eq 'EquipmentMailbox'
```

EXO PS > Set-ManagementRoleAssignment -Identity "OnlyNewCr-GrpOnlyNewCr" `
>> -CustomRecipientWriteScope "EqpCreator"
EXO PS > Get-ManagementRoleAssignment -Identity "OnlyNewCr-GrpOnlyNewCr" |
>> Select-Object RecipientWriteScope,CustomRecipientWriteScope

```
RecipientWriteScope   CustomRecipientWriteScope
-------------------   -------------------------
CustomRecipientScope  EqpCreator
```

等待权限修改生效后，用户 user 01 重新打开一个 PowerShell 的控制台窗口，新创建与 Exchange Online 的连接，并执行以下命令来验证上述修改的结果。修改后完成后，用户 user 01 仅能使用 Get-PublicFolder 命令，且仅可以管理设备邮箱，而非会议室邮箱。命令和输出结果如下：

USER01 PS > Get-PublicFolder

```
Name           Parent Path
----           -----------
IPM_SUBTREE
```

USER01 PS > New-PublicFolder -Name "NewFolder" -Path "\"

```
New-PublicFolder : The term 'New-PublicFolder' is not recognized as the name
of a cmdlet, function, script file, or operable program. Check the spelling of
the name, or if a path was included, verify that the path is correct and try again.
At line:1 char:1
+ New-PublicFolder -Name "NewFolder" -Path "\"
+ ~~~~~~~~~~~~~~~~
    + CategoryInfo          : ObjectNotFound: (New-PublicFolder:String) [],
CommandNotFoundException
    + FullyQualifiedErrorId : CommandNotFoundException
```

USER01 PS > New-Mailbox -Room -Name "CR601"

```
The following error occurred during validation in agent 'Windows LiveId Agent':
'Unable to perform the save
operation.'CHNPR01A001.prod.partner.outlook.cn/Microsoft Exchange Hosted
Organizations/office365ps.partner.onmschina.cn/CR601' is not within a valid
server write scope.'
    + CategoryInfo                    : NotSpecified: (:) [New-Mailbox],
ProvisioningValidationException
    + FullyQualifiedErrorId :
[Server=BJBPR01MB051,RequestId=493a1277-0703-44a8-8550-e1e3535a67ed,TimeStamp
=6/12/2019 7:26:43 AM] [FailureCategory=Cmdlet-ProvisioningValidationException]
E02E993,Microsoft.Exchange.Management.RecipientTasks.NewMailbox
    + PSComputerName          : partner.outlook.cn
```

USER01 PS > New-Mailbox -Equipment -Name "Van05"

```
Name            Alias           Database                        ProhibitSendQuota
ExternalDirectoryObjectId
----            -----           --------                        -----------------
-------------------------
Van05           Van05           CHNPR01DG037-db085              49.5 GB (53,150,2...
8bb0c76d-f813-40b9-b3b4-8f7a48a1...
WARNING: Failed to replicate mailbox:'Van05' to the
site:'CHNPR01.prod.partner.outlook.cn/Configuration/Sites/BJXPR01'.         The
mailbox will be available for logon in approximately 15 minutes.
```

我们也可以从管理员角色组中移除与特定角色相关联的自定义写入作用域，而使该特定角色使用默认的写入作用域。命令和输出结果如下：

EXO PS > Set-ManagementRoleAssignment -Identity "OnlyNewCr-GrpOnlyNewCr" `
>> -CustomRecipientWriteScope $null
EXO PS > Get-ManagementRoleAssignment -Identity "OnlyNewCr-GrpOnlyNewCr" |
>> Select-Object RecipientWriteScope,CustomRecipientWriteScope

```
RecipientWriteScope CustomRecipientWriteScope
------------------- -------------------------
Organization
```

另外，我们还可以使用 Remove-ManagementScope 命令从本租户中彻底删除以前创建的自定义写入作用域。命令和输出结果如下：

EXO PS > Remove-ManagementScope -Identity "CrCreator"

```
Confirm
Are you sure you want to perform this action?
Removing the CrCreator management scope object.
[Y] Yes  [A] Yes to All  [N] No  [L] No to All  [?] Help (default is "Y"): Y
```

（4）我们可以使用 Update-RoleGroupMember 命令更新管理员角色组包括的成员列表，如果仅对个别成员进行增删操作，可以在为 Members 参数赋值的时候，使用 Add 或 Remove 标识符。命令和输出结果如下：

EXO PS > Get-RoleGroupMember -Identity "GrpOnlyNewCr"

```
Name    RecipientType
----    -------------
user01  UserMailbox
user12  UserMailbox
user11  UserMailbox
```

EXO PS > Update-RoleGroupMember -Identity "GrpOnlyNewCr" `
\>> -Members @{Add=(Get-Mailbox -Identity "user 13").Identity;`
\>> Remove=(Get-Mailbox -Identity "user 11").Identity}

```
Confirm
Are you sure you want to perform this action?
Updating the role group "GrpOnlyNewCr" with the member list
"Add=user13;Remove=user11".
[Y] Yes  [A] Yes to All  [N] No  [L] No to All  [?] Help (default is "Y"): A
```

EXO PS > Get-RoleGroupMember -Identity "GrpOnlyNewCr"

```
Name    RecipientType
----    -------------
user01  UserMailbox
user13  UserMailbox
user12  UserMailbox
```

如果我们需要对管理员角色组成员列表中绝大多数的成员进行更新操作，可以直接将新的成员列表赋给 Members 参数。命令和输出结果如下：

EXO PS > Update-RoleGroupMember -Identity "GrpOnlyNewCr" `
\>> -Members (Get-Mailbox -Identity "user 01").Identity,`
\>> (Get-Mailbox -Identity "user 15").Identity,`
\>> (Get-Mailbox -Identity "user 16").Identity

```
Confirm
Are you sure you want to perform this action?
Updating the role group "GrpOnlyNewCr" with the member list "user01, user15, user16".
[Y] Yes  [A] Yes to All  [N] No  [L] No to All  [?] Help (default is "Y"): A
```

EXO PS > Get-RoleGroupMember -Identity "GrpOnlyNewCr"

```
Name    RecipientType
----    -------------
```

```
user01 UserMailbox
user15 UserMailbox
user16 UserMailbox
```

4. 管理员角色组的删除

我们可以使用 Remove-RoleGroup 命令来删除自定义的管理员角色组。命令和输出结果如下：

EXO PS > Remove-RoleGroup -Identity "GrpOnlyNewCr"

```
Confirm
Are you sure you want to perform this action?
Removing the "GrpOnlyNewCr" role group object. The following properties were
configured: management roles "OnlyNewCr", managed by "Organization Management,
 exoAdmin". This operation will also remove the following management role
assignments: OnlyNewCr-GrpOnlyNewCr
[Y] Yes  [A] Yes to All  [N] No  [L] No to All  [?] Help (default is "Y"): Y
```

3.3.2 如何管理用户角色

1. 用户角色策略的查看

在 Exchange 管理中心中单击左侧导航栏中的"权限"按钮，并单击右侧详细内容窗格上方的"用户角色"标签按钮，即可查看本租户内的用户角色策略。另外，我们可以单击用户角色策略列表中特定策略的名称，然后单击策略列表菜单栏中的笔形"编辑"按钮，即可在编辑角色分配策略属性的对话框中查看该用户角色策略的详细属性，如图 3-33 所示。

图 3-33

用户角色策略将授予用户设置其 Outlook 网页版和执行自我管理任务的权限，在默认情况下，每个租户只有单一的默认角色分配策略，且该租户中的所有用户默认都会应用该策略。虽然为了应对企业不同的管理需求，或者为了与企业不同的权限分配原则相匹配，租户中会存在着多个管理员角色组，但是二者在权限授予的管理操作上又是有相似性的。

（1）我们可以使用 Get-RoleAssignmentPolicy 命令来查询本租户默认角色分配策略的属性，并进一步查询其中所包括的 RoleAssignments。类似地，我们也可以使用 ExpandProperty 参数来显示 RoleAssignments 属性的属性值的完整信息。命令和输出结果如下：

```
EXO PS > Get-RoleAssignmentPolicy -Identity "Default Role Assignment Policy" |
>> Select-Object -ExpandProperty RoleAssignments

MyTeamMailboxes-Default Role Assignment Policy
My Custom Apps-Default Role Assignment Policy
My Marketplace Apps-Default Role Assignment Policy
MyMailSubscriptions-Default Role Assignment Policy
MyRetentionPolicies-Default Role Assignment Policy
MyContactInformation-Default Role Assignment Policy
MyDistributionGroupMembership-Default Role Assignment Policy
My ReadWriteMailbox Apps-Default Role Assignment Policy
MyDistributionGroups-Default Role Assignment Policy
MyBaseOptions-Default Role Assignment Policy
MyProfileInformation-Default Role Assignment Policy
MyVoiceMail-Default Role Assignment Policy
MyTextMessaging-Default Role Assignment Policy
```

另外，用户角色策略同样是以 RoleAssignment 为单位进行管理操作的，每个 RoleAssignment 有唯一的名称，命名方式为：用户角色-用户角色策略。我们也可以通过以前使用过的 Get-ManagementRoleAssignment 命令来查询用户角色策略中 RoleAssignment 属性的信息，还可以进一步查询特定 RoleAssignment 的写入作用域等详细信息。命令和输出结果如下：

```
EXO PS > Get-ManagementRoleAssignment `
>> -RoleAssignee "Default Role Assignment Policy"

Name                                Role                    RoleAssigneeName
RoleAssigneeType    AssignmentMethod    EffectiveUserName
----                                ----                    ----------------
----------------    ----------------    -----------------
  MyTeamMailboxes-Default Rol...    MyTeamMailboxes         Default Role Ass...
RoleAssignmentPo... Direct              All Policy Assi...
  My Custom Apps-Default Role...    My Custom Apps          Default Role Ass...
RoleAssignmentPo... Direct              All Policy Assi...
  My Marketplace Apps-Default...    My Marketplace Apps     Default Role Ass...
RoleAssignmentPo... Direct              All Policy Assi...
  MyMailSubscriptions-Default...    MyMailSubscriptions     Default Role Ass...
RoleAssignmentPo... Direct              All Policy Assi...
  MyRetentionPolicies-Default...    MyRetentionPolicies     Default Role Ass...
RoleAssignmentPo... Direct              All Policy Assi...
  MyDistributionGroupMembersh...    MyDistributionGr...     Default Role Ass...
RoleAssignmentPo... Direct              All Policy Assi...
```

```
    My ReadWriteMailbox Apps-De...   My ReadWriteMail...   Default Role Ass...
RoleAssignmentPo... Direct            All Policy Assi...
    MyDistributionGroups-Defaul...   MyDistributionGr...   Default Role Ass...
RoleAssignmentPo... Direct            All Policy Assi...
    MyProfileInformation-Defaul...   MyProfileInforma...   Default Role Ass...
RoleAssignmentPo... Direct            All Policy Assi...
    MyVoiceMail-Default Role As...   MyVoiceMail           Default Role Ass...
RoleAssignmentPo... Direct            All Policy Assi...
    MyTextMessaging-Default Rol...   MyTextMessaging       Default Role Ass...
RoleAssignmentPo... Direct            All Policy Assi...
    MyBaseOptions-Default Ro...      MyBaseOptions         Default Role Ass...
RoleAssignmentPo... Direct            All Policy Assi...
    MyContactInformation-Defaul...   MyContactInforma...   Default Role Ass...
RoleAssignmentPo... Direct            All Policy Assi...
```

EXO PS > Get-ManagementRoleAssignment `
>> -Identity "MyTeamMailboxes-Default Role Assignment Policy" |
>> Select-Object RecipientWriteScope,CustomRecipientWriteScope

```
RecipientWriteScope CustomRecipientWriteScope
------------------- -------------------------
Organization
```

（2）上述以 My 关键字开头的用户角色，授予用户设置其 Outlook 网页版和执行自我管理任务的权限，而这些用户角色又包括用户可以执行的 Exchange Online 命令，我们可以使用 Get-ManagementRoleEntry 命令来进行查询。命令和输出结果如下：

EXO PS > Get-ManagementRoleEntry "MyBaseOptions*"

```
Name                              Role                  Parameters
----                              ----                  ----------
    Set-MailboxMessageConfigura...   MyBaseOptions      {AfterMoveOrDeleteBehavior,
AlwaysShowBcc, AlwaysShowFrom, AutoAddSig...
    ......
    Start-AuditAssistant             MyBaseOptions                  {Identity}
    Set-UserPhoto                    MyBaseOptions                  {Cancel, Confirm,
ErrorAction, ErrorVariable...}
    Set-UnifiedAuditSetting          MyBaseOptions                  {ErrorAction,
ErrorVariable, Identity, OutBuffer...}
    ......
    Search-MessageTrackingReport     MyBaseOptions                  {Confirm,
ErrorAction, ErrorVariable, Identity...}
    Remove-UserPhoto                 MyBaseOptions                  {ClearMailboxPhotoRecord,
Confirm, ErrorAction, ErrorVariable...}
    Remove-SweepRule                 MyBaseOptions                  {Confirm, ErrorAction,
ErrorVariable, Identity...}
    ......
```

```
    Import-ContactList              MyBaseOptions        {Confirm, CSV, CSVData,
CSVStream...}
    Get-UserPhoto                   MyBaseOptions        {Anr, Credential,
ErrorAction, ErrorVariable...}
    Get-User                        MyBaseOptions        {Anr, ErrorAction,
ErrorVariable, Filter...}
    ……
```

2. 用户角色策略属性的修改

我们既可以使用 New-RoleAssignmentPolicy 命令创建新的用户角色策略，并通过 Set-Mailbox 命令将新策略与特定的用户相关联，即授予该用户新策略中的相应权限，也可以通过修改本租户的默认用户角色策略，来调整本租户中所有用户的权限。例如在默认情况下，用户是可以使用下述命令为自己的用户账户设置头像的。命令和输出结果如下：

USER02 PS > Set-UserPhoto -Identity "user02@office365ps.partner.onmschina.cn" `
 >> -PictureData ([System.IO.File]::ReadAllBytes(".\o365.jpg"))

```
Confirm
Are you sure you want to perform this action?
The photo for mailbox user02@office365ps.partner.onmschina.cn will be changed.
[Y] Yes  [A] Yes to All  [N] No  [L] No to All  [?] Help (default is "Y"): Y
```

上述命令成功执行后，当该用户再次登录 Office 365 门户网站的时候，即可在其账户属性中查看该账户头像的图片，如图 3-34 所示。

图 3-34

用户也可以使用下述 Remove-UserPhoto 命令移除自己账户头像的图片。命令和输出结果如下：

USER02 PS > Remove-UserPhoto -Identity "user02@office365ps.partner.onmschina.cn"

```
Confirm
Are you sure you want to perform this action?
The photo for mailbox user02@office365ps.partner.onmschina.cn will be removed.
[Y] Yes  [A] Yes to All  [N] No  [L] No to All  [?] Help (default is "Y"): Y
```

如果公司的 IT 管理策略规定用户不能自行修改 Office 365 账户头像的图片，在 Exchange 管理中心中，无论是通过新建用户角色策略还是修改用户角色策略，都无法满足上述要求；但是，我们可以在 PowerShell 的控制台窗口中通过禁止用户使用与 UserPhoto 相关的 Exchange Online 命令，来解决上述问题。

（1）由于当前生效的用户角色策略当中的多个角色都会包含与 UserPhoto 相关的 Exchange Online 命令，使用户获得执行相关命令的授权。我们可以使用下述命令，查询本租户内有哪些以 My 关键字开头的角色中包含与 UserPhoto 相关的 Exchange Online 命令。命令和输出结果如下：

EXO PS > Get-ManagementRoleEntry -Identity "**UserPhoto" |
>> Where-Object {$_.Role -match "My"}

```
Name                    Role                    Parameters
----                    ----                    ----------
Set-UserPhoto           MyBaseOptions           {Cancel, Confirm, ErrorAction,
ErrorVariable...}
Remove-UserPhoto        MyBaseOptions           {ClearMailboxPhotoRecord,
Confirm, ErrorAction, ErrorVariable...}
Get-UserPhoto           MyBaseOptions           {Anr, Credential, ErrorAction,
ErrorVariable...}
Set-UserPhoto           MyContactInformation    {GroupMailbox, Identity}
Remove-UserPhoto        MyContactInformation    {GroupMailbox, Identity}
Get-UserPhoto           MyContactInformation    {GroupMailbox, Identity}
```

（2）在上述命令的输出结果中，角色 MyBaseOptions 和 MyContactInformation 都包含与 UserPhoto 相关的命令，所以我们以 MyBaseOptions 作为父角色来新建 MyNewBaseOptions 角色，而以 MyContactInformation 作为父角色来创建 MyNewContactInformation 角色。命令和输出结果如下：

EXO PS > New-ManagementRole -Name "MyNewBaseOptions" -Parent "MyBaseOptions"

```
Name                RoleType
----                --------
MyNewBaseOptions    MyBaseOptions
```

```
EXO PS > New-ManagementRole -Name "MyNewContactInformation" `
>> -Parent "MyContactInformation"
```

```
Name                      RoleType
----                      --------
MyNewContactInformation   MyContactInformation
```

（3）在新建自定义角色的时候，我们可以使用 New-ManagementRole 命令配合 EnabledCmdlets 参数来为新创建的角色添加允许用户执行的命令，也可以使用 Remove-ManagementRoleEntry 命令从新创建的角色中移除不允许用户执行的命令。另外，我们还可以移除用户角色中相关命令的参数，以达到限制用户执行特定 Exchange Online 命令的目的。命令和输出结果如下：

```
EXO PS > Set-ManagementRoleEntry `
>> -Identity "MyNewBaseOptions\Remove-UserPhoto" -Parameters $Null
EXO PS > Set-ManagementRoleEntry `
>> -Identity "MyNewBaseOptions\Set-UserPhoto" -Parameters $Null
EXO PS > Set-ManagementRoleEntry `
>> -Identity "MyNewContactInformation\Remove-UserPhoto" -Parameters $Null
EXO PS > Set-ManagementRoleEntry `
>> -Identity "MyNewContactInformation\Set-UserPhoto" -Parameters $Null
EXO PS > Get-ManagementRoleEntry -Identity "*\*UserPhoto" |
>> Where-Object {$_.Role -match "My"}
```

```
Name              Role                        Parameters
----              ----                        ----------
Set-UserPhoto     MyBaseOptions               {Cancel, Confirm, ErrorAction,
ErrorVariable...}
Remove-UserPhoto  MyBaseOptions               {ClearMailboxPhotoRecord,
Confirm, ErrorAction, ErrorVariable...}
Get-UserPhoto     MyBaseOptions               {Anr, Credential,
ErrorAction, ErrorVariable...}
Set-UserPhoto     MyNewBaseOptions            {}
Remove-UserPhoto  MyNewBaseOptions            {}
Get-UserPhoto     MyNewBaseOptions            {Anr, Credential,
ErrorAction, ErrorVariable...}
Set-UserPhoto     MyContactInformation        {GroupMailbox, Identity}
Remove-UserPhoto  MyContactInformation        {GroupMailbox, Identity}
Get-UserPhoto     MyContactInformation        {GroupMailbox, Identity}
Set-UserPhoto     MyNewContactInformation     {}
Remove-UserPhoto  MyNewContactInformation     {}
Get-UserPhoto     MyNewContactInformation     {GroupMailbox, Identity}
```

（4）我们用新创建的 MyNewBaseOptions 角色替换默认用户策略中原有的 MyBaseOptions 角色，并以新创建的 MyNewContactInformation 角色替换默认用户策略中原有的 MyContactInformation 角色。

要注意的是，我们没有新建用户角色策略并与特定的用户账户相关联，而是替换了默认用户策略中的 MyBaseOptions 和 MyContactInformation 角色，即该操作将影响本租户中所有的用户账户。命令和输出结果如下：

EXO PS > Remove-ManagementRoleAssignment `
>> -Identity "MyBaseOptions-Default Role Assignment Policy"

```
Confirm
Are you sure you want to perform this action?
Removing the "MyBaseOptions-Default Role Assignment Policy" management role
assignment object. The following properties were configured: management role
"MyBaseOptions", role assignee "Default Role Assignment Policy", delegation type
"Regular", recipient write scope "Self", and configure write scope
"OrganizationConfig".
 [Y] Yes  [A] Yes to All  [N] No  [L] No to All  [?] Help (default is "Y"): Y
```

EXO PS > New-ManagementRoleAssignment -Role "MyNewBaseOptions" `
>> -Policy "Default Role Assignment Policy"

```
Name                                        Role                   RoleAssigneeName
RoleAssigneeType    AssignmentMethod        EffectiveUserName
----                                        ----                   ----------------
----------------    ----------------        -----------------
 MyNewBaseOptions-Default Ro... MyNewBaseOptions              Default Role Ass...
RoleAssignmentPo... Direct
```

EXO PS > Remove-ManagementRoleAssignment `
>> -Identity "MyContactInformation-Default Role Assignment Policy"

```
Confirm
Are you sure you want to perform this action?
Removing the "MyContactInformation-Default Role Assignment Policy" management
role assignment object. The following properties were configured: management
role "MyContactInformation", role assignee "Default Role Assignment Policy",
delegation type "Regular", recipient write scope "Self", and configure write
scope "OrganizationConfig".
 [Y] Yes  [A] Yes to All  [N] No  [L] No to All  [?] Help (default is "Y"): Y
```

EXO PS > New-ManagementRoleAssignment -Role "MyNewContactInformation" `
>> -Policy "Default Role Assignment Policy"

```
 Name                                      Role                       RoleAssigneeName
RoleAssigneeType        AssignmentMethod   EffectiveUserName
 ----                                      ----                       ----------------
----------------        ----------------   -----------------
 MyNewContactInformation-Def...   MyNewContactInfo...   Default Role Ass...
RoleAssignmentPo...  Direct
```

等待系统刷新完成后，用户 user 02 重新打开一个 PowerShell 的控制台窗口，使用自己的用户凭据创建与 Exchange Online 的连接，并执行以下命令。命令和输出结果如下：

USER02 PS > Set-UserPhoto -Identity "user02@office365ps.partner.onmschina.cn" `
>> -PictureData ([System.IO.File]::ReadAllBytes(".\o365.jpg"))

```
A parameter cannot be found that matches parameter name 'Identity'.
    + CategoryInfo          : InvalidArgument: (:) [Set-UserPhoto],
ParameterBindingException
    + FullyQualifiedErrorId : NamedParameterNotFound,Set-UserPhoto
    + PSComputerName        : partner.outlook.cn
```

USER02 PS > Remove-UserPhoto -Identity "user02@office365ps.partner.onmschina.cn"

```
A parameter cannot be found that matches parameter name 'Identity'.
    + CategoryInfo          : InvalidArgument: (:) [Remove-UserPhoto],
ParameterBindingException
    + FullyQualifiedErrorId : NamedParameterNotFound,Remove-UserPhoto
    + PSComputerName        : partner.outlook.cn
```

在执行上述命令的时候，出现了 A parameter cannot be found that matches parameter name 的报错信息，即用户无法再正常执行 Set-UserPhoto 和 Remove-UserPhoto 命令了。

3.3.3 如何管理 Outlook Web App 策略

1. Outlook Web App 策略的查看

在 Exchange 管理中心中单击左侧导航栏中的"权限"按钮，然后单击右侧详细内容窗格上方的"Outlook Web App 策略"标签按钮，即可查看本租户内的 Outlook Web App 策略。我们也可以单击 Outlook Web App 策略列表中特定策略的名称，接着再单击策略列表菜单栏中的笔形"编辑"按钮，即可在弹出的编辑 Outlook Web App 策略的对话框中查看该策略的详细属性，如图 3-35 所示。

图 3-35

Outlook Web App 策略将为用户 OWA 的可用功能和文件访问应用一套通用的策略，与用户角色策略类似的是，每个租户只有单一的默认策略，且该租户中的所有用户都会应用该默认策略。而与用户角色策略不同的是，Outlook Web App 策略不是以 RoleAssignment 为单位进行管理操作的，且策略中也不嵌套任何的角色。

我们可以使用 Get-OwaMailboxPolicy 命令来查询 Outlook Web App 策略在通信管理方面的属性。命令和输出结果如下：

EXO PS > Get-OwaMailboxPolicy -Identity "OwaMailboxPolicy-Default" |
>> Select-Object InstantMessagingEnabled,TextMessagingEnabled,`
>> UMIntegrationEnabled,ActiveSyncIntegrationEnabled,ContactsEnabled,`
>> AllowCopyContactsToDeviceAddressBook,LinkedInEnabled

```
InstantMessagingEnabled              : True
TextMessagingEnabled                 : True
UMIntegrationEnabled                 : True
ActiveSyncIntegrationEnabled         : True
ContactsEnabled                      : True
AllowCopyContactsToDeviceAddressBook : True
LinkedInEnabled                      : True
```

另外，该命令也可以用来查询 Outlook Web App 策略中与用户体验相关的属性。命令和输出结果如下：

EXO PS > Get-OwaMailboxPolicy -Identity "OwaMailboxPolicy-Default" |
>> Select-Object ThemeSelectionEnabled,PremiumClientEnabled,WeatherEnabled,`
>> PlacesEnabled,LocalEventsEnabled,InterestingCalendarsEnabled

```
ThemeSelectionEnabled                    : True
PremiumClientEnabled                     : True
WeatherEnabled                           : True
PlacesEnabled                            : True
LocalEventsEnabled                       : False
InterestingCalendarsEnabled              : True
```

2. Outlook Web App 策略的修改

请注意，并非所有 OWA 策略的属性都可以在编辑 Outlook Web App 策略的对话框中进行修改，有些属性的修改只能通过使用 PowerShell 的控制台窗口来完成。例如，在保持默认设置的情况下，不具有管理员权限的普通用户也能在 Outlook 网页版中创建 Office 365 组，如图 3-36 所示，而在编辑 Outlook Web App 策略的对话框中却没有可以调整该项设置的开关。

图 3-36

虽然我们无法在编辑 Outlook Web App 策略的对话框中通过修改设置对用户创建 Office 365 组进行限制，但是却可以使用 Set-OwaMailboxPolicy 命令来达到阻止用户在 Outlook 网页版中创建 Office 365 组的目标。命令和输出结果如下：

EXO PS > Set-OwaMailboxPolicy -Identity "OwaMailboxPolicy-Default" `
>> -GroupCreationEnabled $false
EXO PS > Get-OwaMailboxPolicy -Identity "OwaMailboxPolicy-Default" |
>> Select-Object GroupCreationEnabled

```
GroupCreationEnabled
--------------------
           False
```

通过使用 Set-OwaMailboxPolicy 命令来修改 Outlook Web App 策略的属性，有可能需要等待 60 分钟后新的设置才能生效。而在上述修改生效，且用户重新登录 Outlook 网页版后，将无法再看到创建 Office 365 组的"+"按钮。

3. Outlook Web App 策略的创建和删除

我们可以使用 New-OwaMailboxPolicy 命令来新建 Outlook Web App 策略。命令和输出结果如下：

EXO PS > New-OwaMailboxPolicy -Name "NewOwaMailboxPolicy"

```
RunspaceId                          : 399b3de0-ec26-46d5-8a51-5d2e8c74a760
WacEditingEnabled                   : True
PrintWithoutDownloadEnabled         : True
OneDriveAttachmentsEnabled          : True
ThirdPartyFileProvidersEnabled      : False
ClassicAttachmentsEnabled           : True
......
```

另外，我们可以使用 Set-OwaMailboxPolicy 命令配合 IsDefault 参数将指定的 Outlook Web App 策略设置为本租户默认的 OWA 策略，即该策略会影响本租户内的所有用户。要注意的是，我们无法设置 OWA 策略与特定的用户或组相关联，每个租户在任意时刻都只能有一个生效的 OWA 策略，并且租户中当前默认的 OWA 策略是无法删除的。命令和输出结果如下：

EXO PS > Get-OwaMailboxPolicy | Select-Object Name,IsDefault

```
Name                        IsDefault
----                        ---------
NewOwaMailboxPolicy         True
OwaMailboxPolicy-Default    False
```

EXO PS > Remove-OwaMailboxPolicy -Identity "NewOwaMailboxPolicy"

```
The default role assignment policy, NewOwaMailboxPolicy, can't be removed. You
need to set another policy as the default to be able to remove this policy.
    + CategoryInfo                            : WriteError:
(NewOwaMailboxPolicy:OwaMailboxPolicy)        [Remove-OwaMailboxPolicy],
InvalidOperationException
    + FullyQualifiedErrorId :
[Server=BJXPR01MB230,RequestId=cb25c815-7e5b-4f7f-9682-332b1c7c3b1c,TimeStamp
=6/16/2019 3:12:19 AM] [FailureCategory=Cmdlet-InvalidOperationException]
FF389FAA,Microsoft.Exchange.Management.Tasks.RemoveOwaMailboxPolicy
    + PSComputerName         : partner.outlook.cn
```

```
EXO PS > Set-OwaMailboxPolicy -Identity "OwaMailboxPolicy-Default" -IsDefault
```

```
Confirm
Changing the default "OWAMailboxPolicy" to "OwaMailboxPolicy-Default". This
will change the current default policy into a non-default policy. Do you want
to continue?
[Y] Yes  [A] Yes to All  [N] No  [L] No to All  [?] Help (default is "Y"): Y
```

```
EXO PS > Get-OwaMailboxPolicy | Select-Object Name,IsDefault
```

```
Name                        IsDefault
----                        ---------
NewOwaMailboxPolicy         False
OwaMailboxPolicy-Default    True
```

```
EXO PS > Remove-OwaMailboxPolicy -Identity "NewOwaMailboxPolicy"
```

```
Confirm
Are you sure you want to perform this action?
Removing Outlook Web App mailbox policy "NewOwaMailboxPolicy".
[Y] Yes  [A] Yes to All  [N] No  [L] No to All  [?] Help (default is "Y"): Y
```

第 4 章 使用 PowerShell 管理 Office 365 的 Skype for Business Online

Skype for Business Online（以下简称为 Skype for Business）是微软的即时通信工具，它可以实时显示用户的忙闲状态，并提供即时消息、音频和视频在线会议以及桌面演示等功能，使"在家办公"成为可能，但是，对于中小型企业来说，添置服务器硬件，并购买 Skype for Business 的软件许可证，从成本的角度考虑，负担有些大，同时也增加了企业信息服务部门的工作量。而包含在 Office 365 套件中的 Skype for Business 完美地解决了这个问题，使企业有机会以较低的成本拥有这款通讯利器。另外，企业的信息服务部门也不用花费时间和精力来维护硬件设备，可以更多地关注即时通信服务的总体质量。

4.1 如何使用 PowerShell 连接 Skype for Business

使用同一个管理员账号管理企业中所有的系统和服务，存在着安全隐患，同时也无法进行系统的审计；而使用不同的用户凭据管理企业中不同的应用和服务，是我们应该遵循的原则，本章将使用被分配了 Skype for Business 管理员角色的用户 sfb Admin 来进行管理操作。

1. 使用 PowerShell 连接 Skype for Business 的具体步骤

在 Office 365 的管理中心中单击左侧导航栏中的"管理中心"按钮，在弹出的扩展菜单中单击 Skype for Business 按钮，即可进入 Skype for Business 的管理界面，如图 4-1 所示。

图 4-1

因为我们使用的是 sfb Admin 的用户凭据来登录 Office 365 管理中心，而用户 sfb Admin 不是全局管理员，所以只有进入 Skype for Business 管理界面的按钮被显示了出来。也可以使用以下网络地址直接登录 Skype for Business 的管理中心：

https://admin0g.online.partner.lync.cn/lscp

注意

为了能使用 PowerShell 管理 Skype for Business，我们需要使用 64 位版本的操作系统，而且需要安装 Skype for Business 的 Windows PowerShell 模块。

接下来，我们将使用下面的步骤，通过 PowerShell 来连接 Skype for Business。

打开一个 PowerShell 的控制台窗口，键入并执行下面的命令，命令和输出结果如下：

SFB PS > Import-Module "C:\Program Files\Common Files\Skype for Business Online\Modules\SkypeOnlineConnector\SkypeOnlineConnector.psd1"
SFB PS > $credential = Get-Credential

```
cmdlet Get-Credential at command pipeline position 1
Supply values for the following parameters:
Credential
```

SFB PS > $session = New-CsOnlineSession -Credential $credential
SFB PS > Import-PSSession $session

```
ModuleType              Version         Name
ExportedCommands
----------              -------         ----
----------------
Script                  1.0             tmp_mpczgpgw.5rz
{Clear-CsOnlineTelephoneNumberReservation, ConvertTo-JsonForPSWS, Di...
```

与使用 PowerShell 连接 Exchange Online 的方式类似，Skype for Business 使用的也是创建并导入会话的模式。每当我们创建新的连接时，Skype for Business 总是会加载最新的管理模块。这些远程模块的名称，虽然都包括字符串"tmp_"，但是每个模块的名称又各不相同，例如 tmp_mpczgpgw.5rz、tmp_yhfckwc4.cza 等，它们分别代表了不同的会话。

我们可以使用下面的命令查看 Skype for Business 模块包含的 PowerShell 命令的数量。命令和输出结果如下：

SFB PS > $sfbModule = Get-Module -Name "tmp_mpczgpgw.5rz"
SFB PS > $sfbModule.ExportedCommands.Count

393

我们还可以使用 Get-Command 命令配合 Name 参数来获取远程管理模块中所有包含"User"字符串的命令，而这些命令大部分与用户管理的操作有关系。命令和输出结果如下：

SFB PS > Get-Command -Module "tmp_mpczgpgw.5rz" -Name "*User*" |

```
>> Select-Object CommandType,Name,Version

CommandType Name                                                    Version
----------- ----                                                    -------
   Function Disable-CsOnlineDialInConferencingUser                  1.0
   Function Enable-CsOnlineDialInConferencingUser                   1.0
   Function Get-CsExternalUserCommunicationPolicy                   1.0
   Function Get-CsOnlineDialInConferencingUser                      1.0
   Function Get-CsOnlineDialInConferencingUserInfo                  1.0
   Function Get-CsOnlineDialInConferencingUserState                 1.0
   Function Get-CsOnlineUser                                        1.0
   Function Get-CsOnlineVoicemailUserSettings                       1.0
   Function Get-CsOnlineVoiceUser                                   1.0
   Function Get-CsUserAcp                                           1.0
   Function Get-CsUserLocationStatus                                1.0
   Function Get-CsUserPstnSettings                                  1.0
   Function Get-CsUserServicesPolicy                                1.0
   Function Get-CsUserSession                                       1.0
   Function Grant-CsExternalUserCommunicationPolicy                 1.0
   Function Invoke-CsUserPreferredDataLocationSync                  1.0
   Function New-CsExternalUserCommunicationPolicy                   1.0
   Function Remove-CsExternalUserCommunicationPolicy                1.0
   Function Remove-CsUserAcp                                        1.0
   Function Set-CsExternalUserCommunicationPolicy                   1.0
   Function Set-CsOnlineDialInConferencingUser                      1.0
   Function Set-CsOnlineDialInConferencingUserDefaultNumber         1.0
   Function Set-CsOnlineDirectoryUser                               1.0
   Function Set-CsOnlineVoicemailUserSettings                       1.0
   Function Set-CsOnlineVoiceUser                                   1.0
   Function Set-CsOnlineVoiceUserBulk                               1.0
   Function Set-CsUser                                              1.0
   Function Set-CsUserAcp                                           1.0
   Function Set-CsUserPstnSettings                                  1.0
   Function Set-CsUserServicesPolicy                                1.0
```

2. 使用 PowerShell 连接 Skype for Business 的常见问题

（1）超过最大并发连接数量。

与使用 PowerShell 连接 Exchange Online 类似，这种建立永久连接的方式，同一时间只允许数量有限的并行连接。具体的限制是，每个管理员允许创建最多三个到 Skype for Business 的远程连接。如果一个管理员已经有三个正在使用的远程 PowerShell 连接，在其尝试创建第四个连接时，会出现以下错误消息。命令和输出结果如下：

SFB PS (4) > $session = New-CsOnlineSession -Credential $credential

```
New-PSSession : [admin0g.online.partner.lync.cn] Connecting to remote server
```

```
admin0g.online.partner.lync.cn failed with the following error message : The
WS-Management service cannot process the request. The maximum number of concurrent
shells for this user has been exceeded. Close existing shells or raise the quota
for this user. For more information, see the about_Remote_Troubleshooting Help
topic.
    At C:\Program Files\Common Files\Skype for Business
Online\Modules\SkypeOnlineConnector\SkypeOnlineConnectorStartup.psm1:152
    char:16
    + ... $session = New-PSSession -ConnectionUri $ConnectionUri.Uri
-Credenti ...
    +                ~~~~~~~~~~~~~~~~~~~~~~~~~~~~~~~~~~~~~~~~~~~~~~~~~~~~~~~~
        + CategoryInfo                              : OpenError:
(System.Manageme....RemoteRunspace:RemoteRunspace) [New-PSSession],
PSRemotingTransport
        Exception
        + FullyQualifiedErrorId : -2144108123,PSSessionOpenFailed
```

虽然，每个管理员允许最多三个与 Skype for Business 并行连接的会话，但是，所有管理员会话的并行连接数目不能超过 20 个。例如，七个不同的管理员，每个人都创建三个并行的远程连接，而第七个管理员在尝试创建第三个并行连接时将会失败。

解决并行连接数量限制问题的方法也很简单，就是关闭一个或多个以前的会话连接。为了避免长期没有任何操作的会话连接变为孤儿会话，Skype for Business 的管理员最好能够按照最佳操作实践，当不再需要使用一个会话连接时，使用 Remove-PSSession 命令主动地关闭会话。命令和输出结果如下：

SFB PS > Get-PSSession

```
 Id Name            ComputerName        ComputerType       State
ConfigurationName    Availability
 -- ----            ------------        ------------       -----
------------------   ------------
  1 WinRM1          admin0g.onli...     RemoteMachine      Opened
Microsoft.PowerShell Available
```

SFB PS > Remove-PSSession -Session 1

从命令的输出结果我们可以看到，运行 Get-PSSession 命令时，那个 Id 号码就是我们需要赋给 Remove-PSSession 命令中 Session 参数的远程会话连接标识。在不同的 PowerShell 控制台窗口中，如果各自没有多次创建并关闭 Skype for Business 的远程连接的操作，Id 号码始终就是 1。

（2）用户凭据导致的问题。

在使用 PowerShell 连接 Skype for Business 的时候，我们还有可能会遇到以下问题。

如果遇到"登录失败"的问题，我们可以再次检查管理员的用户凭据是否正确。为了帮助我们排查是否由于用户凭据错误导致的登录失败，我们可以尝试使用相同的用户名和密码登录 Office 365 的管理中心：

https://login.partner.microsoftonline.cn

如果出现"用户没有管理此租户权限"的错误，我们应该首先确认该用户是否被分配了 Skype for Business 管理员角色。因为只有 Skype for Business 管理员或本租户的全局管理员才可以建立 Skype for Business 远程会话连接。例如，当我们尝试使用 user 01 的用户凭据创建到 Skype for Business 的远程会话时（user 01 不是 Skype for Business 管理员或全局管理员），会出现下面的报错。命令和输出结果如下：

USER01 PS > $credential = Get-Credential

```
cmdlet Get-Credential at command pipeline position 1
Supply values for the following parameters:
Credential
```

USER01 PS > $session = New-CsOnlineSession -Credential $credential

```
New-PSSession : [admin0g.online.partner.lync.cn] Processing data from remote
server admin0g.online.partner.lync.cn failed with the following error message:
No cmdlets have been authorized for use by the RBAC role that the user belongs
to; therefore a session cannot be established. For more information, see the
about_Remote_Troubleshooting Help topic.
At C:\Program Files\Common Files\Skype for Business
Online\Modules\SkypeOnlineConnector\SkypeOnlineConnectorStartup.psm1:152
    char:16
+ ... $session = New-PSSession -ConnectionUri $ConnectionUri.Uri
-Credenti ...
+                ~~~~~~~~~~~~~~~~~~~~~~~~~~~~~~~~~~~~~~~~~~~~~~~~~~~~~~~~~~
    + CategoryInfo                        : OpenError:
(System.Manageme....RemoteRunspace:RemoteRunspace)      [New-PSSession],
PSRemotingTransport
    Exception
    + FullyQualifiedErrorId : IncorrectProtocolVersion,PSSessionOpenFailed
```

（3）PowerShell 版本过低。

如果由于"PowerShell 版本过低"导致导入模块时出现错误，我们就应该升级 PowerShell 的版本，因为 Skype for Business Connector 模块只能在 PowerShell 3.0 及以上的版本中才能正常地运行。

3. 使用 PowerShell 查询本租户的属性信息

在 Skype for Business 管理中心中单击左侧导航栏中的"仪表板"按钮，在右侧的详细内容窗格中的"组织信息"标签下，即可查看租户的一些基本信息，例如"组织名称""组织 ID"等，如图 4-2 所示。

```
组织信息
下面是您组织的配置信息。 了解详细信息

组织名称：                          PS for Office 365
组织 ID：                           6578fce9-2a78-4b8f-be6f-a5916a19dd49
组织创建时间：                      2018/4/2 19:17:56
Active Directory 同步对象：         否
组织域：                            office365ps.partner.onmschina.cn
```

图 4-2

使用 Get-CsTenant 命令，并结合使用 PowerShell 的管道操作符对命令结果进行过滤，我们可以得到与 Skype for Business 管理中心界面格式类似的该租户的属性信息。命令和输出结果如下：

SFB PS > Get-CsTenant |
\>> Select-Object DisplayName,TenantId,WhenCreated,DirSyncEnabled,Domains

```
DisplayName      : PS for Office 365
TenantId         : 6578fce9-2a78-4b8f-be6f-a5916a19dd49
WhenCreated      : 4/3/2018 10:17:56 AM
DirSyncEnabled   :
Domains          : {office365ps.partner.onmschina.cn}
```

对比后我们发现，使用 Get-CsTenant 命令得到的组织创建时间 WhenCreated 和 Skype for Business 管理中心的仪表板上显示的"组织创建时间"是有差异的，而导致差异的原因在于 PowerShell 命令显示的是北京时间，而管理中心显示的是美国太平洋时间。

通过使用 Get-CsTenant 命令，我们还可以查询到比 Skype for Business 管理中心更为详细的租户信息，例如租户订阅信息和租户标识信息等。以下这些信息对 Skype for Business 的管理员和 Office 365 的技术支持工程师排错也是有帮助的。命令和输出结果如下：

SFB PS > Get-CsTenant | Select-Object Identity,AssignedPlan,`
\>> ProvisionedPlan,UserRoutingGroupIds,LastSyncTimeStamp

```
Identity        : OU=6578fce9-2a78-4b8f-be6f-a5916a19dd49,OU=OCS Tenants,DC=
                  lync0g001,DC=local
AssignedPlan   :{<XmlValueAssignedPlan xmlns:xsd="http://www.w3.org/2001/
                  XMLSchema"
                    xmlns:xsi="http://www.w3.org/2001/XMLSchema-instance">
                    <Plan
SubscribedPlanId="2f7f38c7-3e9c-4cce-ae02-be6957265787"
                    ServiceInstance="MicrosoftCommunicationsOnline/CN-0G-S"
CapabilityStatus="Enabled"
                    AssignedTimestamp="2018-04-03T02:16:42Z" ServicePlanId=
"0feaeb32-d00e-4d66-bd5a-43b5b83db82c"
                    xmlns="http://schemas.microsoft.com/online/
                    directoryservices/change/2008/11">
```

```
                    <Capability>
                      <Capability Plan="MCOProfessional"
xmlns="http://schemas.microsoft.com/online/MCO/2009/01" />
                    </Capability>
                  </Plan>
                </XmlValueAssignedPlan>}
ProvisionedPlan     : {<XmlValueProvisionedPlan
                    xmlns:xsd="http://www.w3.org/2001/XMLSchema"
                    xmlns:xsi="http://www.w3.org/2001/XMLSchema-instance">
                    <Plan SubscribedPlanId="2f7f38c7-3e9c-4cce-ae02-be6957265787"
                       ServiceInstance="MicrosoftCommunicationsOnline/CN-0G-S
CapabilityStatus="Enabled"

                       AssignedTimestamp="2018-04-03T02:16:42Z" ProvisioningStatus=
"Success"
                       ProvisionedTimestamp="2018-04-03T02:18:00.4692792Z"
                    xmlns="http://schemas.microsoft.com/online/
                    directoryservices/change/2008/11" />
                  </XmlValueProvisionedPlan>}
UserRoutingGroupIds : {196a3ddf-8b65-55bf-8e63-83e07c7d9fa0}
LastSyncTimeStamp   : 4/29/2019 6:37:15 PM
```

4.2 如何使用 PowerShell 管理 Skype for Business 的用户

我们可以使用 Get-CsOnlineUser 命令来查询 Skype for Business 用户的常规或外部通信等属性，使用 Grant-CsConferencingPolicy 和 Grant-CsExternalAccessPolicy 等命令分别修改用户的会议和外部通信等以策略方式定义的属性。

4.2.1 如何查看用户

1. 用户的查看

在 Skype for Business 管理中心中单击左侧导航栏中的"用户"按钮，在右侧的详细内容窗格中即可查看 Skype for Business 的用户列表，如图 4-3 所示。

图 4-3

我们可以使用 Get-CsOnlineUser 命令显示与 Skype for Business 管理中心类似的用户列表。命令和输出结果如下：

```
SFB PS > Get-CsOnlineUser |
>> Where-Object {$_.InterpretedUserType -eq "PureOnlineSfBUser"} |
>> Select-Object DisplayName,UserPrincipalName,UsageLocation |
>> Sort-Object DisplayName

DisplayName     UserPrincipalName                              UsageLocation
-----------     -----------------                              -------------
exo Admin       exoAdmin@office365ps.partner.onmschina.cn      CN
FrankCai        admin@office365ps.partner.onmschina.cn         CN
sfb Admin       sfbadmin@office365ps.partner.onmschina.cn      CN
spo Admin       spoadmin@office365ps.partner.onmschina.cn      CN
user 01         user01@office365ps.partner.onmschina.cn        CN
user 02         user02@office365ps.partner.onmschina.cn        CN
……
```

我们将 InterpretedUserType 的属性值设置为 PureOnlineSfBUser，是为了使结果集仅包括本租户内部的用户，而不包括其他类型的用户对象，例如外部用户等。使用 Get-CsOnlineUser 命令还可以查询 Skype for Business 管理中心的用户列表中无法显示、却对管理员有参考价值的属性，例如 Identity、Guid、ProvisionedPlan 等。另外，我们还可以使用 ResultSize 参数来限制结果集中返回结果的数量。

2. 用户的筛选

在用户列表的菜单栏中，如果没有"搜索"文本框，可以单击放大镜形状的"搜索"按钮使其弹出。而通过在搜索文本框中键入搜索关键字，并按"回车"键，我们可以对用户列表进行筛选和过滤，如图 4-4 所示。

图 4-4

我们可以键入部分显示名称作为搜索字符串来进行模糊查询，需要注意的是，在搜索文本框中键入的搜索关键字不能包含"*"通配符，否则将无法显示正确的结果。类似地，使用 Get-CsOnlineUser 命令配合 Identity 参数和通配符，我们同样也可以按照关键字查询用户。命令和输出结果如下：

```
SFB PS > Get-CsOnlineUser -Identity "*er1*" |
>> Where-Object {$_.InterpretedUserType -eq "PureOnlineSfBUser"} |
>> Select-Object DisplayName,UserPrincipalName,UsageLocation |
>> Sort-Object DisplayName

DisplayName   UserPrincipalName                        UsageLocation
-----------   -----------------                        -------------
user 11       user11@office365ps.partner.onmschina.cn  CN
user 12       user12@office365ps.partner.onmschina.cn  CN
user 13       user13@office365ps.partner.onmschina.cn  CN
user 15       user15@office365ps.partner.onmschina.cn  CN
user 16       user16@office365ps.partner.onmschina.cn  CN
user 18       user18@office365ps.partner.onmschina.cn  CN
```

4.2.2 如何查看用户的属性

在 Skype for Business 管理中心中单击左侧导航栏中的"用户"按钮，在右侧的用户列表中，选择用户显示名称左侧的复选框，然后在页面右侧弹出的该用户的属性对话框中单击笔形的"编辑"按钮，如图 4-5 所示，即可在新打开的页面中查看用户的"常规"或"外部通信"等设置。

图 4-5

1. 用户会议策略属性的查看

（1）在新打开的用户属性页面中单击左侧导航栏中的"常规"按钮，即可在右侧的详细内容窗格中查看该用户常规选项的设置（用户属性页面中的常规选项在PowerShell中对应于用户会议策略这个属性），如图4-6所示。

图 4-6

用户属性页面中常规项下的音频、视频、记录会话和会议等选项的设置，与 PowerShell 中用户的 ConferencingPolicy 属性的属性值相对应，我们可以通过使用 Get-CsOnlineUser 命令来查询用户该属性的属性值。命令和输出结果如下：

SFB PS > Get-CsOnlineUser -Identity "user 16" |
>> Select-Object DisplayName, ConferencingPolicy

```
DisplayName ConferencingPolicy
----------- ------------------
user 16     BposSAllModality
```

（2）常规项下的选项包括是否允许用户使用音频和视频、是否记录对话和会议、是否关闭非存档功能等内容。而针对以上各项设置所产生的不同选项组合，ConferencingPolicy 属性均有唯一的属性值与之对应，请参考表 4-1。

表 4-1

音频和视频	记录对话和会议	为确保合规性，请关闭非存档功能	属性值
音频和 HD 视频	选择	（空白）	BposSAllModality（默认值）
	选择	选择	BposSAllModalityNoFT
	（空白）	选择	BposSDataProtection
	（空白）	（空白）	BposSAllModalityNoRec
音频和视频	选择	（空白）	BposSAllModalityMinVideoBW
	选择	选择	BposSAllModalityNoFTMinVideoBW
	（空白）	选择	BposSDataProtectionMinVideoBW
	（空白）	（空白）	BposSAllModalityNoRecMinVideoBW
仅音频	选择	（空白）	BposSAllModalityNoVideo
	选择	选择	BposSAllModalityNoFTNoVideo
	（空白）	选择	BposSDataProtectionNoVideo
	（空白）	（空白）	BposSAllModalityNoRecNoVideo

(续表)

音频和视频	记录对话和会议	为确保合规性,请关闭非存档功能	属性值
无	选择	(空白)	BposSVoipDisabled
	选择	选择	BposSIMPOnly
	(空白)	选择	BposSIMPOnlyNoRec
	(空白)	(空白)	BposSVoipDisabledNoRec

上述 ConferencingPolicy 属性的属性值,即各种会议策略的名称标签,可以使用下述 Get-CsConferencingPolicy 命令进行查询。命令和输出结果如下:

SFB PS > Get-CsConferencingPolicy | Select-Object Identity

```
WARNING: Showing only policies that are applicable to your current
subscription(s). To see other policies, try "-Include All"
Identity
--------
Global
Tag:BposSAllModality
Tag:BposSAllModalityNoVideo
Tag:BposSAllModalityMinVideoBW
Tag:BposSAllModalityNoDialout
Tag:BposSAllModalityNoDialoutNoVideo
Tag:BposSAllModalityNoDialoutMinVideoBW
Tag:BposSAllModalityNoFT
Tag:BposSAllModalityNoFTNoVideo
Tag:BposSAllModalityNoFTMinVideoBW
Tag:BposSAllModalityNoFTNoDialout
Tag:BposSAllModalityNoFTNoDialoutNoVideo
Tag:BposSAllModalityNoFTNoDialoutMinVideoBW
Tag:BposSAllModalityNoRec
Tag:BposSAllModalityNoRecNoVideo
Tag:BposSAllModalityNoRecMinVideoBW
Tag:BposSAllModalityNoRecNoDialout
Tag:BposSAllModalityNoRecNoDialoutNoVideo
Tag:BposSAllModalityNoRecNoDialoutMinVideoBW
Tag:BposSDataProtection
Tag:BposSDataProtectionNoVideo
Tag:BposSDataProtectionMinVideoBW
Tag:BposSDataProtectionNoDialout
Tag:BposSDataProtectionNoDialoutNoVideo
Tag:BposSDataProtectionNoDialoutMinVideoBW
Tag:BposSIMPOnly
Tag:BposSIMPOnlyNoDialout
Tag:BposSIMPOnlyNoRec
Tag:BposSIMPOnlyNoRecNoDialout
```

```
Tag:BposSVoipDisabled
Tag:BposSVoipDisabledNoDialout
Tag:BposSVoipDisabledNoRec
Tag:BposSVoipDisabledNoRecNoDialout
Tag:BposSAllModalityWithPPANoFT
Tag:BposSVoipDisabledWithPPANoFT
```

另外，我们还可以使用该命令查询具体会议策略的各项子属性，并通过比较不同策略的子属性的属性值来进一步了解不同策略间的差异。例如，策略 BposSAllModality 允许用户使用音频和 HD 视频功能；而策略 BposSAllModalityNoVideo 仅允许用户使用音频功能，在视频功能方面的属性处于关闭的状态。命令和输出结果如下：

SFB PS > Get-CsConferencingPolicy -Identity "BposSAllModality" |
>> Select-Object Identity,AllowIPAudio,AllowIPVideo,AllowMultiView,EnableP2PVideo

```
Identity         : Tag:BposSAllModality
AllowIPAudio     : True
AllowIPVideo     : True
AllowMultiView   : True
EnableP2PVideo   : True
```

SFB PS > Get-CsConferencingPolicy -Identity "BposSAllModalityNoVideo" |
>> Select-Object Identity,AllowIPAudio,AllowIPVideo,AllowMultiView,EnableP2PVideo

```
Identity         : Tag:BposSAllModalityNoVideo
AllowIPAudio     : True
AllowIPVideo     : False
AllowMultiView   : False
EnableP2PVideo   : False
```

2. 用户外部通信属性的查看

（1）在用户属性页面中单击左侧导航栏中的"外部通信"按钮，即可在右侧的窗格中查看该用户外部通信选项的设置，如图 4-7 所示。

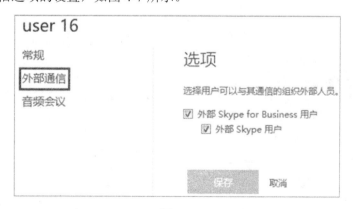

图 4-7

类似地，用户属性页面中外部通信项下的设置，与 PowerShell 中用户的 ExternalAccessPolicy 属性的属性值相对应，我们同样可以通过使用 Get-CsOnlineUser 命令来获取用户该属性的属性值。命令和输出结果如下：

```
SFB PS > Get-CsOnlineUser -Identity "user 16" |
>> Select-Object DisplayName, ExternalAccessPolicy

DisplayName ExternalAccessPolicy
----------- --------------------
user 16     FederationAndPICDefault
```

（2）外部通信选项的设置决定了用户是否可以与外部的 Skype for Business 用户或外部的 Skype 用户进行通信。而针对以上各项设置所产生的不同种有效组合，ExternalAccessPolicy 属性均有唯一的属性值与之对应，请参考表 4-2。

表 4-2

外部 Skype for Business 用户	外部 Skype 用户	属性值
选择	选择	FederationAndPICDefault（默认值）
选择	（空白）	FederationOnly
（空白）	（空白）	NoFederationAndPIC
（空白）	选择	（不受支持的类型）

上述 ExternalAccessPolicy 属性的属性值，即各种外部通信策略的名称标签，我们可以使用下述 Get-CsExternalAccessPolicy 命令进行查询。命令和输出结果如下：

```
SFB PS > Get-CsExternalAccessPolicy | Select-Object Identity

WARNING: Showing only policies that are applicable to your current
subscription(s). To see other policies, try "-Include All"
Identity
--------
Global
Tag:FederationAndPICDefault
Tag:FederationOnly
Tag:NoFederationAndPIC
```

而且，我们也可以使用该命令查询具体外部通信策略的各子属性。

```
SFB PS > Get-CsExternalAccessPolicy -Identity "FederationAndPICDefault"

Identity                : Tag:FederationAndPICDefault
Description             :
EnableFederationAccess  : True
EnableXmppAccess        : False
EnablePublicCloudAccess : True
```

```
EnablePublicCloudAudioVideoAccess      : True
EnableOutsideAccess                    : True
```

3. 用户音频会议属性的查看

在用户属性页面中单击左侧导航栏中的"音频会议"按钮，然后在右侧的窗格中查看该用户音频会议属性的设置，如图 4-8 所示。

图 4-8

在图形界面中，需要打开每个用户的用户属性页面来查看音频会议项下的设置；而使用 PowerShell 的 Get-CsOnlineUser 命令或者 Get-CsUserAcp 命令，则可以方便地查询特定用户的音频会议提供商的属性信息。命令和输出结果如下：

SFB PS > Get-CsOnlineUser -Identity "user 16" | Select-Object DisplayName,AcpInfo

```
DisplayName AcpInfo
----------- -------
user 16     {}
```

SFB PS > Get-CsUserAcp -Identity "user 16" | Select-Object Name,AcpInfo

```
Name                                    AcpInfo
----                                    -------
9f0b20d5-1f28-4cfe-bc98-166452f51b3e   {}
```

4. 其他用户属性的查看

（1）我们仅可以在用户属性页面中查看常规选项、外部通信选项和音频会议选项的设置，还有一些用户属性，我们是无法在图形界面上进行查看的。对于这些图形界面上无法查看的属性，我们依然可以使用 Get-CsOnlineUser 命令来进行查询。例如我们可以使用下述命令查询用户 user 16 的客户端策略。命令和输出结果如下：

```
SFB PS > Get-CsOnlineUser -Identity "user 16" | Select-Object DisplayName,ClientPolicy

DisplayName ClientPolicy
----------- ------------
user 16
```

在命令执行的结果中，ClientPolicy 属性的属性值为空，即用户 user 16 应用的是该属性的默认全局属性值。

（2）我们可以使用 Get-CsClientPolicy 命令来查询客户端策略的名称标签、特定的客户端策略中的各项子属性及它们对应的属性值。命令和输出结果如下：

```
SFB PS > Get-CsClientPolicy | Select-Object Identity

Identity
--------
Global
Tag:ClientPolicyNoSaveIMNoArchivingNoIMURL
Tag:ClientPolicyNoIMURL
Tag:ClientPolicyNoSaveIMNoArchiving
Tag:ClientPolicyNoSaveIMNoArchivingNoIMURLPhoto
Tag:ClientPolicyNoIMURLPhoto
Tag:ClientPolicyNoSaveIMNoArchivingPhoto
Tag:ClientPolicyDefault
Tag:ClientPolicyDefaultPhoto
Tag:ClientPolicyDisableSkypeUI
Tag:ClientPolicyEnableSkypeUI
Tag:ClientPolicyNoSaveIMNoArchivingNoIMURLDisableSkypeUI
Tag:ClientPolicyNoIMURLDisableSkypeUI
Tag:ClientPolicyNoSaveIMNoArchivingDisableSkypeUI
Tag:ClientPolicyDefaultPhotoDisableSkypeUI
Tag:ClientPolicyNoSaveIMNoArchivingNoIMURLPhotoDisableSkypeUI
Tag:ClientPolicyNoIMURLPhotoDisableSkypeUI
Tag:ClientPolicyNoSaveIMNoArchivingPhotoDisableSkypeUI
Tag:CPCPstnPreviewRateMyCallClientPolicy
```

```
SFB PS > Get-CsClientPolicy -Identity "Global" |
>> Select-Object Identity,DisableEmoticons, DisableInkIM

Identity DisableEmoticons DisableInkIM
-------- ---------------- ------------
Global
```

5. 查看用户属性涉及的两个问题

（1）在 Skype for Business 管理中心的用户列表中，单击用户显示名称左侧的复选框以选择该用户，在页面右侧弹出的用户属性对话框中，即可查看该用户常规和外部通信属性的各项属性值。然而，如果我们此时再复选另一个用户，且这两个用户在同一属性上的设置并不完全相同,就会导致用户属性对话框将该属性从属性列表中移除,即不再显示该属性。以上这个特性意味着,Skype for Business 的管理员无法在 Skype for Business 管理中心的用户列表中同时查看分属多个用户的不相同的属性值,例如,用户 user 13 仅可以使用 Skype for Business 的音频功能,而 user 15 可以使用音频和 HD 视频功能,同时选择这两个用户将使用用户属性对话框不再显示常规属性项,如图 4-9 所示。

图 4-9

而使用 Get-CsOnlineUser 命令,却可以在单一的结果列表中,以自定义的顺序,同时显示多个属性及用户在这些属性上对应的属性值。命令和输出结果如下:

```
SFB PS > Get-CsOnlineUser -Identity "*ser1*" |
>> Where-Object {$_.InterpretedUserType -eq "PureOnlineSfBUser"} |
>> Select-Object DisplayName,ConferencingPolicy,ExternalAccessPolicy |
>> Sort-Object DisplayName
```

```
DisplayName ConferencingPolicy              ExternalAccessPolicy
----------- ------------------              --------------------
user 11     BposSAllModalityNoFTNoVideo     FederationAndPICDefault
user 12     BposSAllModality                NoFederationAndPIC
user 13     BposSAllModalityNoRecNoVideo    FederationAndPICDefault
user 15     BposSDataProtection             FederationAndPICDefault
user 16     BposSAllModality                FederationAndPICDefault
user 18     BposSAllModality                FederationOnly
```

(2) 在 Skype for Business 管理中心用户列表的菜单栏中如果没有"查看用户"下拉列表框,我们可以单击漏斗形状的"查看用户"按钮使其弹出。"查看用户"下拉列表框包括了一系列的视图名称,用于检索设置了相应的常规或外部通信属性的用户,如图 4-10 所示。

第 4 章 使用 PowerShell 管理 Office 365 的 Skype for Business Online

图 4-10

使用 Get-CsOnlineUser 命令配合 Filter 参数，我们可以达到与应用筛选视图类似的过滤效果。

注意

可能会出现单个视图的筛选条件与特定属性的多个属性值相匹配的情况，我们此时就需要检查每种可能出现的情况，以获取完整的用户列表。

例如，只要用户属性页面常规选项中的"记录对话和会议"选项未被选择，即该用户不保存对话和会议记录，该用户就应该出现在"关闭录制"视图的结果列表中，而与此种情况相匹配的 ConferencingPolicy 属性值总共有 8 个，我们需要逐一对它们进行检查。命令和输出结果如下：

```
SFB PS > Get-CsOnlineUser -Filter {(ConferencingPolicy -eq "BposSDataProtection") `
>> -or (ConferencingPolicy -eq "BposSAllModalityNoRec") `
>> -or (ConferencingPolicy -eq "BposSData ProtectionMinVideoBW") `
>> -or (ConferencingPolicy -eq "BposSAllModalityNoRecMinVideoBW") `
>> -or (ConferencingPolicy -eq "BposSDataProtectionNoVideo") `
>> -or (ConferencingPolicy -eq "BposSAllModalityNoRecNoVideo") `
>> -or (ConferencingPolicy -eq "BposSIMPOnlyNoRec") `
>> -or (ConferencingPolicy -eq "BposSVoipDisabledNoRec")} |
>> Where-Object{$_.InterpretedUserType -eq "PureOnlineSfBUser"} |
>> Select-Object DisplayName,UserPrincipalName,UsageLocation |
>> Sort-Object DisplayName

DisplayName  UserPrincipalName                         UsageLocation
-----------  -----------------                         -------------
user 13      user13@office365ps.partner.onmschina.cn   CN
user 15      user15@office365ps.partner.onmschina.cn   CN
```

4.2.3 如何修改用户的属性

在 Skype for Business 管理中心中单击左侧导航栏中的 "用户" 按钮；然后，在右侧的用户列表中选择用户显示名称左侧的复选框；接着，在页面右侧弹出的用户属性对话框中单击笔形的 "编辑" 按钮，请参考图 4-5；最后，在新出现的页面中修改用户的 "常规" 或 "外部通信" 等属性，请参考图 4-6 和图 4-7。使用 PowerShell 对用户的常规、外部通信等属性进行修改时会涉及多个不同的命令。

> **注意**
> 虽然我们可以使用 Get-CsOnlineUser 命令来查询 Skype for Business 用户的各项属性，但是要注意，不存在用来修改用户属性的 Set-CsOnlineUser 命令。

1. 用户会议策略属性的修改

（1）PowerShell 中的用户会议策略属性，对应于 Skype for Business 管理中心用户属性页面的常规选项。我们可以使用 Grant-CsConferencingPolicy 命令将会议策略应用到特定的用户，来达到修改该用户会议策略属性的目的。命令和输出结果如下：

SFB PS > Grant-CsConferencingPolicy -PolicyName "Tag:BposSAllModalityNoVideo" `
\>> -Identity "user16@office365ps.partner.onmschina.cn"
SFB PS > Get-CsOnlineUser -Identity "user 16" |
\>> Select-Object DisplayName, ConferencingPolicy

```
DisplayName  ConferencingPolicy
-----------  ------------------
user 16      BposSAllModalityNoVideo
```

（2）我们可以使用 New-CsConferencingPolicy 命令来新建会议策略，自定义新策略中的各项子属性，然后将该新策略应用到特定的用户。并且，我们还可以利用 Set-CsConferencingPolicy 命令来修改自定义会议策略中的子属性，比如 AllowAnonymous ParticipantsInMeetings 子属性。而这些属性在 Skype for Business 管理中心用户属性页面的常规选项中是无法进行修改的。命令和输出结果如下：

SFB PS > New-CsConferencingPolicy -Identity "NoAllowAnonymous" `
\>> -AllowAnonymousParticipantsInMeetings $false

```
Identity                                : Tag:NoAllowAnonymous
……
AllowAnonymousUsersToDialOut            : False
AllowAnonymousParticipantsInMeetings    : False
……
```

SFB PS > Grant-CsConferencingPolicy -PolicyName "Tag:NoAllowAnonymous" `
\>> -Identity "user18@office365ps.partner.onmschina.cn"

```
SFB PS > Get-CsOnlineUser -Identity "user 18" |
>> Select-Object DisplayName, ConferencingPolicy
```

```
DisplayName  ConferencingPolicy
-----------  ------------------
user 18      NoAllowAnonymous
```

上述命令新创建了 NoAllowAnonymous 会议策略，并给用户 user 18 应用了该会议策略。该会议策略不允许匿名用户加入会议，即匿名用户将无法加入用户 user 18 发起的会议。

（3）虽然我们可以使用 Remove-CsConferencingPolicy 命令删除自定义的会议策略，但是当前与用户相关联的自定义会议策略及 Skype for Business 预定义的会议策略是无法被删除的。命令和输出结果如下：

```
SFB PS > Remove-CsConferencingPolicy -Identity "Tag:ConfinedExtUser"
SFB PS > Remove-CsConferencingPolicy -Identity "Tag:NoAllowAnonymous"
```

```
The policy "NoAllowAnonymous" is currently assigned to one or more users. Assign
a different policy to the users before removing this one.
    + CategoryInfo          : InvalidOperation: (NoAllowAnonymous:String)
[Remove-CsConferencingPolicy], InvalidOperationException
    + FullyQualifiedErrorId :
Error,Microsoft.Rtc.Management.Internal.RemoveHostedConferencingPolicyCmdlet
    + PSComputerName        : admin0g.online.partner.lync.cn
```

```
SFB PS > Remove-CsConferencingPolicy -Identity "Tag:BposSVoipDisabledNoRec"
```

```
Set or Remove host tag instance is not allowed
    + CategoryInfo          : NotSpecified: (:) [Remove-CsConferencingPolicy],
ArgumentException
    + FullyQualifiedErrorId :
System.ArgumentException,Microsoft.Rtc.Management.Internal.RemoveHostedConfer
encingPolicyCmdlet
    + PSComputerName        : admin0g.online.partner.lync.cn
```

我们尝试删除的上述会议策略中，ConfinedExtUser 和 NoAllowAnonymous 是自定义的会议策略，而 BposSVoipDisabledNoRec 是 Skype for Business 预定义的会议策略。

2. 用户外部通信属性的修改

（1）我们可以使用 Grant-CsExternalAccessPolicy 命令将外部通信策略应用到特定的用户，来达到修改该用户外部通信属性的目的。命令和输出结果如下：

```
SFB PS > Grant-CsExternalAccessPolicy -PolicyName "Tag:FederationOnly" `
>> -Identity "user16@office365ps.partner.onmschina.cn"
SFB PS > Get-CsOnlineUser -Identity "user 16" |
>> Select-Object DisplayName, ExternalAccessPolicy
```

```
DisplayName   ExternalAccessPolicy
-----------   --------------------
user 16       FederationOnly
```

（2）我们可以使用 New-CsExternalAccessPolicy 命令来新建外部通信策略，自定义新策略中的各项子属性，并将该策略应用到特定的用户；也可以利用 Set-CsExternal-AccessPolicy 命令来修改自定义外部通信策略中的属性。这些属性在 Skype for Business 管理中心用户属性页面的外部通信选项中也是无法进行修改的。另外，我们还可以使用 Remove-CsExternalAccessPolicy 命令删除自定义的外部通信策略。

3. 用户音频会议属性的修改

我们可以使用 Set-CsUserAcp 命令来修改特定用户的音频会议属性。命令必选的参数包括 Identity、TollNumber、ParticipantPasscode、Domain 和 Name，其中除了 Identity 参数外其他参数都和音频会议的提供商有关。命令和输出结果如下：

SFB PS > Set-CsUserAcp -Identity "user16@office365ps.partner.onmschina.cn" `
>> -TollNumber "8314265858" -ParticipantPasscode "5858" -Domain "polycom.com.cn" `
>> -Name "Polycom"

```
WARNING: The 'Url' option is not specified or an empty string is specified
when creating a new user ACP Info record, which may result in 'Find a local
number' link missing in the meeting invitation.
```

SFB PS > Get-CsOnlineUser -Identity "user 16" | Select-Object -ExpandProperty AcpInfo

```
<acpInformation>
  <tollNumber>8314265858</tollNumber>
  <participantPassCode>5858</participantPassCode>
  <domain>polycom.com.cn</domain>
  <name>Polycom</name>
</acpInformation>
```

4. 其他用户属性的修改

（1）与修改用户的 ConferencingPolicy 属性和 ExternalAccessPolicy 属性的属性值类似，我们为用户修改 MobilityPolicy 和 PresencePolicy 等属性的时候，需要分别使用 Grant-CsMobilityPolicy 命令和 Grant-CsPresencePolicy 命令。

（2）在这些属性中，ClientPolicy 属性是经常被修改的属性之一，为用户应用该属性中的策略决定了该用户可以使用 Skype for Business 应用中的哪些功能。命令和输出结果如下：

SFB PS > Grant-CsClientPolicy -PolicyName "Tag:ClientPolicyDefaultPhoto" `
>> -Identity "user16@office365ps.partner.onmschina.cn"
SFB PS > Get-CsOnlineUser -Identity "user 16" | Select-Object DisplayName,ClientPolicy

```
DisplayName  ClientPolicy
-----------  ------------
user 16      ClientPolicyDefaultPhoto
```

（3）因为 ClientPolicy 属性包括默认定义的全局策略，所以我们不能再新建全局策略，而只能使用 Set-CsClientPolicy 命令修改默认的全局策略，但是，我们可以使用 New-CsClientPolicy 命令新建自定义的客户端策略，并将该新建客户端策略与特定的用户相关联。例如，下述命令将新建客户端策略 NewClientPolicy，该新策略将禁止用户收发表情符号和含数字墨水技术的平板电脑手写会话内容。命令和输出结果如下：

SFB PS > New-CsClientPolicy -Identity "Tag:NewClientPolicy" -DisableEmoticons $true `
\>\> -DisableInkIM $true

```
Identity                                  : Tag:NewClientPolicy
PolicyEntry                               :
{Name=OnlineFeedbackUrl;Value=http://go.microsoft.com/fwlink/?LinkID=224932,
Name=EnableTraceRouteReporting;Value=TRUE,
Name=AllowAdalForNonLyncIndependentOfLync;Value=TRUE}
Description                               :
AddressBookAvailability                   : WebSearchOnly
AttendantSafeTransfer                     :
AutoDiscoveryRetryInterval                :
BlockConversationFromFederatedContacts    :
……
```

SFB PS > Grant-CsClientPolicy -PolicyName "Tag:NewClientPolicy" `
\>\> -Identity"user18@office365ps.partner.onmschina.cn"
SFB PS > Get-CsOnlineUser -Identity "user 18" | Select-Object DisplayName,ClientPolicy

```
DisplayName  ClientPolicy
-----------  ------------
user 18      NewClientPolicy
```

另外，我们可以使用 Remove-CsClientPolicy 命令删除自定义的客户端策略。

4.3 如何使用 PowerShell 管理 Skype for Business 的组织

与管理单个用户不同，管理 Skype for Business 的组织是对本租户的会议策略或外部通信等属性进行配置或修改等操作，将影响本租户中的所有用户。尤其是对外部通信属性的配置或修改，将影响本租户与其他域的联盟设置，即本租户中的用户是否可以与其他域中的用户建立通信联系。

4.3.1 如何管理"状态隐私模式"属性

设置状态隐私模式是为了管控是否允许租户内的其他人查看用户的状态信息，用户的状态信息除了包括忙、闲、在线或离线等，还包括用户的位置信息及头像等。

1. "状态隐私模式"属性的查看

在 Skype for Business 管理中心左侧的导航栏中单击"组织"按钮，在右侧详细内容窗格"常规"标签下的"状态隐私模式"项中可以查看或修改"状态隐私模式"的设置，如图 4-11 所示。

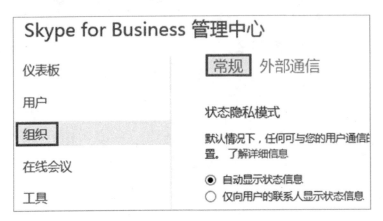

图 4-11

使用 Get-CsPrivacyConfiguration 命令可以查看本租户的"状态隐私模式"的设置，在默认情况下，EnablePrivacyMode 属性的值是 False，也就是租户内的用户相互之间可以查看状态信息。命令和输出结果如下：

SFB PS > Get-CsPrivacyConfiguration

```
Identity                        : Global
EnablePrivacyMode               : False
AutoInitiateContacts            : True
PublishLocationDataDefault      : True
DisplayPublishedPhotoDefault    : True
```

2. "状态隐私模式"属性的修改

如果将 EnablePrivacyMode 属性的值设置成 True，则只有用户的联系人才能看到该用户的状态信息，而成为用户联系人的必要步骤是，用户将该人加入自己的联系人列表。通过使用 Set-CsPrivacyConfiguration 命令可以对状态隐私模式的属性值进行修改。命令和输出结果如下：

SFB PS > Set-CsPrivacyConfiguration -EnablePrivacyMode $true

```
WARNING: Because you are enabling privacy mode, we recommend that you also
enable client version check in order to block communications with client
versions that do not recognize privacy mode. We also recommend that you set
AutoInitiateContacts, PublishLocationDataDefault, and
DisplayPublishedPhotoDefault to false. When privacy mode is enabled, Response
Group agents must add presence watchers to their contact list, or else the
watchers will not be able to see the agents' presence.
```

运行以上命令的时候，会出现警告信息。提示我们最好将 PublishLocationDataDefault 和 DisplayPublishedPhotoDefault 两个属性的值设为 False，这样可以隐藏用户的位置信息和头像，并在用户明确要求展示时才会进行展示，以进一步加强用户隐私的保护。另外，警告信息还提示我们将 AutoInitiateContacts 值也设置为 False，这样可以避免用户的上级和下属自动被加入用户的联系人列表。上述这三个关于"状态隐私模式"的属性值只能通过 PowerShell 的命令进行设置，而无法在图形界面常规标签的状态隐私模式项下进行修改。

> **注意**
>
> New-CsPrivacyConfiguration 和 Remove-CsPrivacyConfiguration 两条命令只能在本地版本的 Skype for Business 服务器上才能被使用。

4.3.2 如何管理"移动电话通知"属性

启用移动电话通知服务是为了让用户在他们电脑上面的 Skype for Business 客户端处于挂起状态时，能在 Windows Phone 或苹果的 iPhone 上接收到 Skype for Business 消息的推送，推送的内容包括即时消息、会议邀请等，前提是用户已经在手机或平板电脑上安装了 Skype for Business 的客户端应用。

1. "移动电话通知"属性的查看

我们可以在 Skype for Business 管理中心左侧的导航栏中单击"组织"按钮，并在右侧详细内容窗格的"常规"标签下的"移动电话通知"项中查看或修改该设置，如图 4-12 所示。

图 4-12

推送通知服务目前只支持 Microsoft 推送通知服务和 Apple 推送服务，而这两项服务默认都是处于关闭状态的。可以使用 Get-CsPushNotificationConfiguration 命令查看本租户该属性的设置情况。命令和输出结果如下：

SFB PS > Get-CsPushNotificationConfiguration

```
Identity                                  : Global
EnableApplePushNotificationService        : False
EnableMicrosoftPushNotificationService    : False
```

2．"移动电话通知"属性的修改

使用 Set-CsPushNotificationConfiguration 命令可以分别修改 Microsoft 推送通知服务和 Apple 推送通知服务的状态。下面的命令将启用本租户的 Microsoft 和 Apple 推送通知服务。命令如下：

SFB PS > Set-CsPushNotificationConfiguration `
>> -EnableMicrosoftPushNotificationService $true `
>> -EnableApplePushNotificationService $true

类似地，New-CsPushNotificationConfiguration 和 Remove-CsPushNotificationConfiguration 两条命令只能在本地版本的 Skype for Business 服务器上才能使用。

4.3.3 如何管理"外部通信"属性

1．"外部通信"属性的查看

通过为租户设置 Skype for Business 的联盟，可以允许内部用户与特定的外部域中的用户交换会话初始协议（SIP）消息，并且可以与这些外部域中的联系人互通状态信息。

在 Skype for Business 管理中心左侧的导航栏中单击"组织"按钮，然后在右侧详细内容窗格的"外部通信"标签下，即可查看"外部访问"和"公共 IM 连接"设置，如图 4-13 所示。

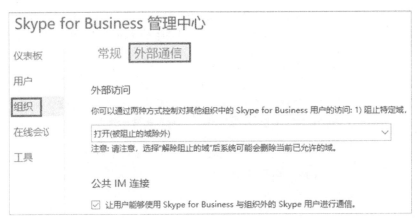

图 4-13

（1）我们可以使用下面的 PowerShell 命令来查询本租户 Skype for Business 联盟的设置情况。命令和输出结果如下：

SFB PS > Get-CsTenantFederationConfiguration

```
Identity                                :Global
AllowedDomains                          :AllowAllKnownDomains
BlockedDomains                          :{}
AllowFederatedUsers                     :True
AllowPublicUsers                        :True
TreatDiscoveredPartnersAsUnverified     :False
SharedSipAddressSpace                   :False
```

从结果中我们可以看到，在默认设置下，本租户中的用户允许与所有外部域中的用户进行通信，并且允许与公共即时通信系统中的用户进行通信，即 AllowFederatedUsers 和 AllowPublicUsers 这两个属性的值均为 True，且 AllowedDomains 属性的值为 AllowAllKnownDomains，而 BlockedDomains 属性的值为空。

（2）在图形界面的"外部通信"标签项下，我们是无法查看 Skype for Business 支持与哪些公共的即时通信系统中的用户进行通信的，但是我们却可以使用 PowerShell 的 Get-CsTenantPublicProvider 命令来进行查询，并得到结果。命令和输出结果如下：

SFB PS > Get-CsTenantPublicProvider

```
RunspaceId          :43915e91-2be3-4e93-aac8-e027b8e1fe4d
PublicProviderSet   :True
DomainPICStatus     :{Microsoft.Rtc.Management.Hosted.DomainPICStatus,
Microsoft.Rtc.Management.Hosted.DomainPICStatus,Microsoft.Rtc.Management.Host
ed.DomainPICStatus}
```

结果中 PublicProviderSet 属性的值是 True，意味着本租户至少与一个公共即时通信系统的供应商建立了联盟。我们可以使用下面的命令查看，与各个公共即时通信系统供应商建立联盟的具体情况。命令和输出结果如下：

SFB PS > Get-CsTenantPublicProvider -Tenant (Get-CsTenant).TenantId |
>> Select-Object -ExpandProperty DomainPICStatus

```
Provider    :WindowsLive
Domain      :office365ps.partner.onmschina.cn
Status      :Enabled
TimeStamp   :5/4/2019 2:56:40 PM
Detail      :

Provider    :AOL
Domain      :office365ps.partner.onmschina.cn
Status      :Disabled
TimeStamp   :1/1/0001 12:00:00 AM
Detail      :


Provider    :Yahoo
Domain      :office365ps.partner.onmschina.cn
```

```
Status          :Disabled
TimeStamp       :1/1/0001 12:00:00 AM
Detail          :
```

上面的输出结果表明本租户与公共即时通信系统的供应商 WindowsLive 建立了联盟，如果本租户的策略又允许本租户中的用户与公共即时通信系统中的用户进行通信，用户便可以与 WindowsLive 系统中的存在账户的用户进行通信了。

2. "外部通信"属性的修改

（1）我们可以使用下面的命令，让本租户既能与公共即时通信系统的供应商 WindowsLive，也能与供应商 AOL，建立联盟关系。命令如下：

SFB PS > Set-CsTenantPublicProvider -Tenant (Get-CsTenant).TenantId `
>> -Provider "WindowsLive","AOL"

接下来，使用下面的命令查看修改后的结果。命令和输出结果如下：

SFB PS > Get-CsTenantPublicProvider -Tenant (Get-CsTenant).TenantId |
>> Select-Object -ExpandProperty DomainPICStatus

```
Provider        : WindowsLive
Domain          : office365ps.partner.onmschina.cn
Status          : Enabled
TimeStamp       : 5/4/2019 2:56:40 PM
Detail          :

Provider        : AOL
Domain          : office365ps.partner.onmschina.cn
Status          : PendingEnabled
TimeStamp       : 5/20/2019 11:44:39 PM
Detail          :

Provider        : Yahoo
Domain          : office365ps.partner.onmschina.cn
Status          : Disabled
TimeStamp       : 1/1/0001 12:00:00 AM
Detail          :
```

注意

调整本租户与公共即时通信系统供应商的联盟设置需要一些时间，而在设置修改的过程中，状态将会被显示为 PendingEnabled 或 PendingDisabled。

如果要禁止本租户中的用户与公共即时通信系统中的用户进行通信，我们可以使用 Set-CsTenantFederationConfiguration 命令。当以下命令成功执行后，即使 Get-CsTenantPublicProvider 命令显示本租户允许与一个或多个公共即时通信系统的供应商建立联盟，用户都将无法再与所有的公共即时通信系统中的存在账户的用户进行通信了。命令如下：

SFB PS > Set-CsTenantFederationConfiguration -AllowPublicUsers $False

（2）如果要与租户中的其他用户联系，在不修改系统默认设置的情况下，只要搜索用户邮箱或用户名便可以在 Skype for Business 客户端上看到对方的状态，并可以与其创建会话，如图 4-14 所示。

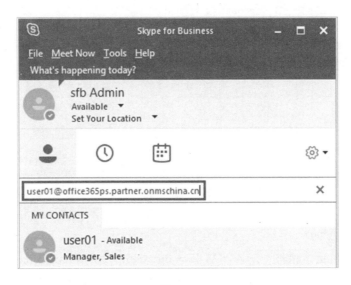

图 4-14

同样使用 Skype for Business 的默认设置，本租户中的用户是允许与其他租户中的用户进行通信的，前提条件是其他租户也没有修改外部通信的默认设置，或者允许与本租户建立 Skype for Business 联盟。要注意的是，当尝试与其他租户中的用户进行通信时，必须使用对方完整邮件地址形式的用户名，如图 4-15 所示。

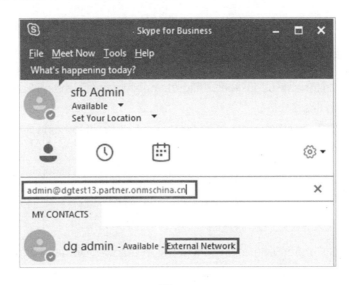

图 4-15

例外情况是，当本租户中的用户已经将其他租户中的用户添加到通讯录或者已经与其有过通信的记录，则该用户仅使用对方的用户名也可以查看对方的状态或创建新的会话。另外，其他租户中的用户在 Skype for Business 客户端中的显示条目都包含 External Network 的标识符。

本租户可以与其他哪些租户建立 Skype for Business 的联盟，或者说本租户中的用户具体可以与其他哪些租户中的用户进行通信，是域名黑白名单相互作用的结果。我们可以使用下面的命令查看域名的黑白名单，命令和输出结果如下：

```
SFB PS > Get-CsTenantFederationConfiguration |
>> Select-Object AllowedDomains, BlockedDomains

AllowedDomains          BlockedDomains
--------------          --------------
AllowAllKnownDomains {}
```

更加具体一些，白名单可以使用两种类型的属性值，第一类属性值是 AllowAllKnownDomains，也就是允许与所有的外部域建立联盟，此时在黑名单中出现的域名将被禁止与其建立联盟，即本租户不会与这些出现在黑名单中的租户建立联盟关系。第二类白名单的属性值是一组域名的列表，而黑名单在这种情况下会被忽略，本租户仅允许与白名单中列出的域名建立联盟。

例如，我们要阻止本租户中的用户与来自 dgtest13.partner.onmschina.cn 域的用户进行通信，需要首先创建一个包含所有域的白名单。由于 Skype for Business 不允许我们在 PowerShell 命令中使用字符串直接更新黑白名单列表，我们就需要使用下面的这条命令创建一个包含所有域的白名单对象（如果没有改变过默认设置，下面的命令可以省略。）命令如下：

```
SFB PS > $allowed = New-CsEdgeAllowAllKnownDomains
```

接下来，我们使用下面的这条命令创建一个黑名单列表对象。命令如下：

```
SFB PS > $blocked = New-CsEdgeDomainPattern -Domain "dgtest13.partner.onmschina.cn"
```

最后，我们使用 Set-CsTenantFederationConfiguration 命令来更新黑白名单。命令如下：

```
SFB PS > Set-CsTenantFederationConfiguration -AllowedDomains $allowed `
>> -BlockedDomains $null
SFB PS > Set-CsTenantFederationConfiguration -AllowedDomains $allowed `
>> -BlockedDomains @{Add=$blocked}
```

上述第一条命令将原有黑名单的属性值设为$null，即删除黑名单中的所有条目，恢复到空的状态。第二条命令使用参数值@{Add=$blocked}添加只包括一个域名的黑名单对象。而下面的这条命令与上述两条命令是等效的。命令如下：

```
SFB PS > Set-CsTenantFederationConfiguration -AllowedDomains $allowed `
>> -BlockedDomains @{Replace = $blocked}
```

如果要创建包含部分域名的白名单列表对象，我们可以使用 New-CsEdgeAllowList 命令。修改完成后，我们再次查看外部通信的设置。命令和输出结果如下：

SFB PS > Get-CsTenantFederationConfiguration

```
Identity                                    : Global
AllowedDomains                              : AllowAllKnownDomains
BlockedDomains                              : {Domain=dgtest13.partner.onmschina.cn}
AllowFederatedUsers                         : True
AllowPublicUsers                            : True
TreatDiscoveredPartnersAsUnverified         : False
SharedSipAddressSpace                       : False
```

上述域的黑白名单设置生效后，在 Skype for Business 的客户端中，来自 dgtest13.partner.onmschina.cn 域的用户的状态将变为 Presence unknown，即我们无法再查看该域中用户的状态并与其进行通信了，如图 4-16 所示。

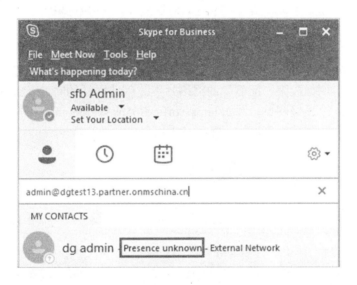

图 4-16

另外要再提醒一下，对外部通信设置的修改要等待一段时间后才会生效，特别是在解除对某个域的访问限制后，用户并不能立即查询到该域中用户的状态。

第 5 章 使用 PowerShell 管理 Office 365 的 SharePoint Online

SharePoint Portal Server 是微软的智能门户站点解决方案，构建在一个可伸缩的架构之上，包括了内容管理和业务流程管理的服务，同时提供了用于部署、开发和管理的工具包，让用户可以拥有定制化和个性化的门户站点体验。然而，部署在本地的 SharePoint Portal Server 需要管理员进行的日常工作包括维护底层的硬件设施、监控服务的健康状态及进行备份等，这会使系统管理员长期处于疲于奔命的状态。而通过使用 Office 365 的 SharePoint Online，系统管理员可以把更多的精力放在网站管理、用户权限分配等工作上面，进而为协同工作这一信息化目标提供更好的支持。

SharePoint Online 中的网站包括顶级网站和其下的子网站。网站是一组具有相同所有者，并且共享管理设置，例如权限、配额等的网络站点。创建网站的时候，SharePoint Online 会自动在网站中创建一个顶级网站，例如 https://office365ps.sharepoint.cn。随后，我们可以在顶级网站的下面创建一个或多个子网站，如图 5-1 所示。

图 5-1

虽然在默认情况下，顶级网站和子网站具有相同的所有者且共享管理设置；但是，我们可以在不同的层级上对顶级网站或子网站的功能和设置进行控制。例如，我们可以让公司内的不同部门使用同一网站中的不同子网站，或者为不同的部门创建单独的网站。另外，我们对网站的规划安排应该取决于公司的组织结构及业务规模。如果我们能预先了解一些基本的信息，例如，如何使用网站、哪些用户将有访问权限等，就能更好地帮助我们选择网站的模板，或者进行存储空间的分配。

> **注意**
>
> 本书中用到的"网站"和"站点"具有相同的含义，"顶级网站"和"根网站"是对同一对象的不同称谓。另外，经典 SharePoint 管理中心的"网站集"和新版管理中心的"网站"

第 5 章 使用 PowerShell 管理 Office 365 的 SharePoint Online

这两个称谓，如果没有特殊说明，在一般情况下是可以相互替换的。

目前，SharePoint Online 的管理中心处于新旧版本的更替过程中，习惯经典页面的用户可以在新版的 SharePoint 管理中心中单击左侧导航栏中的"设置"按钮，在右侧的"设置"窗格内单击"默认管理中心"按钮，然后在页面右侧弹出的"默认管理中心"对话框中，选择使用"经典 SharePoint 管理中心"。

5.1 如何使用 PowerShell 连接 SharePoint Online

本章绝大多数的操作都会使用 spo Admin 的用户凭据进行，而我们已经为该用户分配了 SharePoint Online 管理员的角色。

1. 使用 PowerShell 连接 SharePoint Online 的具体步骤

在 Office 365 管理中心中单击左侧导航栏中的"管理中心"按钮，在弹出的扩展菜单中单击 SharePoint 按钮，如图 5-2 所示，进入 SharePoint Online 的管理中心。

图 5-2

SharePoint Online 管理中心也可以使用以下网址直接登录：

https://office365ps-admin.sharepoint.cn/_layouts/15/online/AdminHome.aspx#/home

如果要使用 PowerShell 连接 SharePoint Online，请确保 SharePoint Online Management Shell 模块已经安装。如果该模块已经成功安装，我们只需要在 Windows 系统的开始菜单中单击 SharePoint Online Management Shell 的快捷方式将其打开即可。当然，我们也可以打开一个 PowerShell 的控制台窗口，然后键入以下 PowerShell 命令加载 SharePoint Online Management Shell 的模块。命令如下：

SPO PS > Import-Module Microsoft.Online.SharePoint.PowerShell -DisableNameChecking

上述加载 SharePoint 模块的步骤并不是必需的，因为从 PowerShell 的 3.0 版本开始，当执行 Connect-SPOService 命令的时候，便会自动加载所需的模块了。

通过单击快捷方式打开的 SharePoint Online Management Shell 的控制台窗口看起来与默认的 PowerShell 控制台窗口类似，不同点是通过快捷方式打开的控制台的标题为 SharePoint Online Management Shell，并且窗口的默认底色是黑色的，如图 5-3 所示。

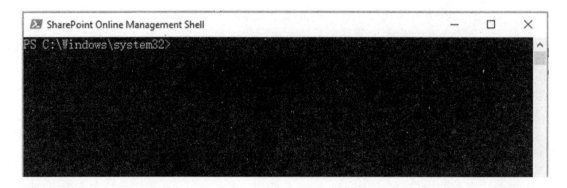

图 5-3

我们可以运行以下命令来连接 SharePoint Online。命令和输出结果如下：

SPO PS > $cred= Get-Credential

```
cmdlet Get-Credential at command pipeline position 1
Supply values for the following parameters:
Credential
```

SPO PS > $uri = 'https://office365ps-admin.sharepoint.cn'
SPO PS > Connect-SPOService -Url $uri -Credential $cred

在弹出的登录窗口中键入用户 spo Admin 的账号密码，然后单击"确定"按钮，验证成功之后，我们将返回到 SharePoint Online 的控制台窗口，此时就可以键入 SharePoint Online 的命令了。例如，我们可以列出 SharePoint Online Management Shell 中可用命令的数量。命令和输出结果如下：

SPO PS > $spocmds = Get-Command |
>> Where-Object {$_.ModuleName -eq "Microsoft.Online.SharePoint.PowerShell"}
SPO PS > $spocmds.Count

186

上面的第一条命令将 SharePoint Online Management Shell 管理模块中所有的命令保存到 $spocmds 变量中，第二条命令输出对$spocmds 变量计数的结果。从命令执行的结果可知，目前有 186 条命令可以用于管理 SharePoint Online，而命令的数量也会随着 SharePoint Online 的更新发生变化。

2. 使用 PowerShell 连接 SharePoint Online 的常见问题

与 Skype for Business Online 类似，只有为用户分配了 SharePoint Online 管理员的角色，或者使用本租户的全局管理员用户凭据，才可以建立与 SharePoint Online 的远程会话连接。如果我们尝试使用 user 01 的用户凭据建立与 SharePoint Online 的远程会话连接时（user 01 不是 SharePoint Online 的管理员），将会出现以下报错信息。命令和输出结果如下：

```
USER01 PS > $cred= Get-Credential
```

```
cmdlet Get-Credential at command pipeline position 1
Supply values for the following parameters:
Credential
```

```
USER01 PS > $uri = 'https://office365ps-admin.sharepoint.cn'
USER01 PS > Connect-SPOService -Url $uri -Credential $cred
```

```
Connect-SPOService : The remote server returned an error: (401) Unauthorized.
At line:1 char:1
+ Connect-SPOService -Url $uri -Credential $cred
+ ~~~~~~~~~~~~~~~~~~~~~~~~~~~~~~~~~~~~~~~~~~~~~~
    + CategoryInfo          : NotSpecified: (:) [Connect-SPOService],
WebException
    + FullyQualifiedErrorId :
System.Net.WebException,Microsoft.Online.SharePoint.PowerShell.ConnectSPOServ
ice
```

3. 使用 PowerShell 查询本租户的属性信息

SharePoint Online 是以多租户的模式运行的，可以使用 Get-SPOTenant 命令获取本租户的详细信息，还可以使用管道命令过滤输出结果，更加迅速地找到我们关注的属性。命令和输出结果如下：

```
SPO PS > Get-SPOTenant
```

```
StorageQuota                                : 1560576
StorageQuotaAllocated                       : 340787200
ResourceQuota                               : 10300
ResourceQuotaAllocated                      : 300
OneDriveStorageQuota                        : 1048576
CompatibilityRange                          : 15,15
ExternalServicesEnabled                     : False
NoAccessRedirectUrl                         :
SharingCapability                           : ExternalUserSharingOnly
DisplayStartASiteOption                     : True
StartASiteFormUrl                           :
ShowEveryoneClaim                           : False
ShowAllUsersClaim                           : False
OfficeClientADALDisabled                    : False
LegacyAuthProtocolsEnabled                  : True
ShowEveryoneExceptExternalUsersClaim        : True
SearchResolveExactEmailOrUPN                : False
RequireAcceptingAccountMatchInvitedAccount  : True
ProvisionSharedWithEveryoneFolder           : False
SignInAccelerationDomain                    :
```

```
EnableGuestSignInAcceleration               : False
UsePersistentCookiesForExplorerView         : False
BccExternalSharingInvitations               : False
BccExternalSharingInvitationsList           :
UserVoiceForFeedbackEnabled                 : True
PublicCdnEnabled                            : False
PublicCdnAllowedFileTypes                   : CSS,EOT,GIF,ICO,JPEG,JPG,
                                              JS,MAP,PNG,SVG,TTF,WOFF
PublicCdnOrigins                            : {}
RequireAnonymousLinksExpireInDays           : -1
SharingAllowedDomainList                    :
SharingBlockedDomainList                    :
SharingDomainRestrictionMode                : None
OneDriveForGuestsEnabled                    : False
IPAddressEnforcement                        : False
IPAddressAllowList                          :
IPAddressWACTokenLifetime                   : 15
UseFindPeopleInPeoplePicker                 : False
DefaultSharingLinkType                      : Internal
ODBMembersCanShare                          : Unspecified
ODBAccessRequests                           : Unspecified
PreventExternalUsersFromResharing           : False
ShowPeoplePickerSuggestionsForGuestUsers    : False
FileAnonymousLinkType                       : Edit
FolderAnonymousLinkType                     : Edit
NotifyOwnersWhenItemsReshared               : True
NotifyOwnersWhenInvitationsAccepted         : True
NotificationsInOneDriveForBusinessEnabled   : True
NotificationsInSharePointEnabled            : True
SpecialCharactersStateInFileFolderNames     : Allowed
OwnerAnonymousNotification                  : True
CommentsOnSitePagesDisabled                 : False
SocialBarOnSitePagesDisabled                : False
OrphanedPersonalSitesRetentionPeriod        : 30
PermissiveBrowserFileHandlingOverride       : False
DisallowInfectedFileDownload                : False
DefaultLinkPermission                       : View
ConditionalAccessPolicy                     : AllowFullAccess
AllowDownloadingNonWebViewableFiles         : True
AllowEditing                                : True
ApplyAppEnforcedRestrictionsToAdHocRecipients : True
FilePickerExternalImageSearchEnabled        : True
EmailAttestationRequired                    : False
EmailAttestationReAuthDays                  : 30
```

5.2 如何使用 PowerShell 管理 SharePoint Online 的网站

SharePoint Online Management Shell 中管理网站的命令，可以按照添加和删除网站、设置网站等功能进行细分。管理 SharePoint Online 网站的 PowerShell 命令中总是包含 SPOSite 这个标识符，SPO 是 SharePoint Online 的缩写形式，而管理本地版本的 SharePoint 网站的命令中会包含 SPSite 标识符。

5.2.1 如何查看网站与修改网站的属性

1. 活动站点及其属性的查看

（1）在 SharePoint 管理中心左侧的导航栏中单击"网站"按钮，并在弹出的扩展菜单中单击"活动站点"按钮，在右侧窗格的活动站点列表中即可查看本租户内所有的网站，如图 5-4 所示。

图 5-4

相应地，我们可以使用不带任何参数的 Get-SPOSite 命令查看本租户内所有的网站。命令和输出结果如下：

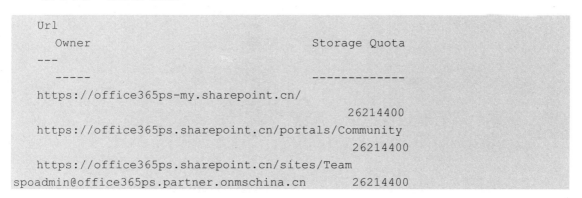

```
       https://office365ps.sharepoint.cn/portals/hub
                                                26214400
       https://office365ps.sharepoint.cn/sites/notemplate
spoadmin@office365ps.partner.onmschina.cn       26214400
       https://office365ps.sharepoint.cn/sites/Docs
spoadmin@office365ps.partner.onmschina.cn       26214400
       https://office365ps.sharepoint.cn/
                                                26214400
       https://office365ps.sharepoint.cn/sites/Entwiki
spoadmin@office365ps.partner.onmschina.cn       26214400
       https://office365ps.sharepoint.cn/sites/Records
spoadmin@office365ps.partner.onmschina.cn       26214400
       https://office365ps.sharepoint.cn/sites/Projects
spoadmin@office365ps.partner.onmschina.cn       26214400
       https://office365ps.sharepoint.cn/search
                                                26214400
```

注意

上述结果中包括以团队网站为模板创建的网站，不包括个人网站的子站点。如果结果中不包括以团队网站为模板创建的网站，我们可以在控制面板的程序和功能列表中查看 SharePoint Online Management Shell 安装包的版本信息，并确保该安装包的版本至少是"16.0.8924.1200"。

（2）在 SharePoint 管理中心的活动站点列表中单击特定网站的站点名称，即可在页面右侧弹出的网站属性对话框中查看该网站的属性，如图 5-5 所示。

Site Office365PS

常规　活动　权限　策略

网站名称
Site Office365PS
编辑

URL
https://office365ps.sharepoint.cn
编辑

中心关联
无

存储空间上限
25.00 TB
编辑

模板
团队网站(经典体验)

Office 365 组已连接
否

域
office365ps.sharepoint.cn

说明
无

图 5-5

而使用 Get-SPOsite 命令并配合 Identity 参数，我们则可以查询到比在 SharePoint 管理中心中更为详细的网站属性信息，例如 CompatibilityLevel、LockState 及 Status 等。命令和输出结果如下：

SPO PS > Get-SPOSite -Identity "https://office365ps.sharepoint.cn" | Format-List

```
LastContentModifiedDate                      : 5/11/2019 8:22:53 PM
Status                                       : Active
ResourceUsageCurrent                         : 0
ResourceUsageAverage                         : 0
StorageUsageCurrent                          : 2
LockIssue                                    :
WebsCount                                    : 1
CompatibilityLevel                           : 15
DisableSharingForNonOwnersStatus             :
HubSiteId                                    : 00000000-0000-0000-0000-000000000000
IsHubSite                                    : False
Url                                          : https://office365ps.sharepoint.cn/
LocaleId                                     : 2052
LockState                                    : Unlock
Owner                                        : eb09cf2a-6c70-4e5b-ad82-8cc01c8b97ab
StorageQuota                                 : 26214400
StorageQuotaWarningLevel                     : 25574400
ResourceQuota                                : 50
ResourceQuotaWarningLevel                    : 0
Template                                     : STS#0
Title                                        : PS for Office 365 工作组网站
AllowSelfServiceUpgrade                      : True
DenyAddAndCustomizePages                     : Disabled
PWAEnabled                                   : Unknown
SharingCapability                            : ExternalUserSharingOnly
SiteDefinedSharingCapability                 : ExternalUserAndGuestSharing
SandboxedCodeActivationCapability            : Disabled
DisableCompanyWideSharingLinks               : NotDisabled
DisableAppViews                              : NotDisabled
DisableFlows                                 : NotDisabled
StorageQuotaType                             :
RestrictedToGeo                              : Unknown
ShowPeoplePickerSuggestionsForGuestUsers     : False
SharingDomainRestrictionMode                 : None
SharingAllowedDomainList                     :
SharingBlockedDomainList                     :
ConditionalAccessPolicy                      : AllowFullAccess
AllowDownloadingNonWebViewableFiles          : False
AllowEditing                                 : True
CommentsOnSitePagesDisabled                  : False
SocialBarOnSitePagesDisabled                 : False
```

（3）在默认情况下，如果我们最初使用的是经典 SharePoint 管理中心，并且购买了"Office 365 商业高级版"的订阅，在经典 SharePoint 管理中心的网站集列表中，我们便可以查看 SharePoint Online 默认创建的三个网站集。表 5-1 总结了这三个默认网站集的链接地址及网站集的模板 ID。

表 5-1

网 站 集	网站集链接	网站集模板 ID
（根）网站	https://office365ps.sharepoint.cn	EHS#1
搜索中心	https://office365ps.sharepoint.cn/search	SRCHCEN#0
我的网站	https://office365ps-my.sharepoint.cn	SPSMSITEHOST#0

注意

在新版 SharePoint 管理中心中，我们无法在活动站点的列表中查看"我的网站"和"搜索中心"这两个网站，而只能通过使用 PowerShell 的 Get-SPOsite 命令对这两个网站进行查询。

2. 网站属性的修改

我们可以使用 Set-SPOSite 命令来修改网站的属性。例如，下述命令将网站的标题修改为"Office365PS Site"。命令如下：

```
SPO PS > Set-SPOSite -Identity "https://office365ps.sharepoint.cn" `
>> -Title "Office365PS Site"
```

另外，我们还可以使用该命令配合不同的参数来修改网站的其他属性，该命令支持的参数请参考表 5-2。

表 5-2

参数名称	作用描述
AllowSelfServiceUpgrade	设置是否允许网站管理员升级其网站
DenyAddAndCustomizePages	设置是否允许用户自定义网站上的页面选项
LocaleId	设置网站所使用语言的区域 ID
LockState	设置网站是否被锁定
Owner	设置网站的所有者
ResourceQuota	设置网站的资源配额
ResourceQuotaWarningLevel	设置网站资源配额的警告级别
SandboxedCodeActivationCapability	设置是否为网站启用或禁用沙盒解决方案
StorageQuota	设置网站的存储配额（以兆字节为单位）
StorageQuotaWarningLevel	设置网站存储配额的警告级别（以兆字节表示）
Titile	设置网站的标题

5.2.2 如何添加与删除网站

1. 网站的添加

（1）在 SharePoint 管理中心左侧的导航栏中单击"网站"按钮，然后在弹出的扩展菜单中单击"活动站点"按钮，并在右侧窗格的活动站点列表上方的菜单栏中单击"创建"按钮，如图 5-6 所示；接着，在页面右侧弹出的"创建网站"对话框中单击"团队网站"、"通信网站"或"其他选项"按钮，来新建网站。

图 5-6

使用 PowerShell 命令 New-SPOSite 创建新的 SharePoint Online 网站时，必须要添加三个参数：网站的 URL、网站所有者的用户标识符和存储配额。网站的模板 ID 可以作为 Template 参数的属性值添加，但并不是强制性的。如果省略该参数，SharePoint Online 将会强制网站的管理员在第一次访问新创建的网站的时候，从"选择模板"列表中选择需要使用的模板，如图 5-7 所示。

图 5-7

使用下述的 PowerShell 命令，我们可以创建一个以商业智能中心为模板的网站。命令如下：

SPO PS > New-SPOSite -URL "https://office365ps.sharepoint.cn/sites/BI" `
>> -Owner "spoadmin@office365ps.partner.onmschina.cn" -StorageQuota 500 `
>> -Template BICenterSite#0

另外，我们可以创建开发人员网站，为开发人员使用该网站来开发自定义的业务应用程序并在云端进行测试提供方便。使用下述 PowerShell 命令，通常只需要很短的时间就可以创建一个全新的开发人员网站。命令如下：

SPO PS > New-SPOSite -URL "https://office365ps.sharepoint.cn/sites/dev" `
>> -Owner "spoadmin@office365ps.partner.onmschina.cn" -StorageQuota 500 `
>> -Template DEV#0

（2）使用命令 Get-SPOWebTemplate 可以获取 SharePoint Online 提供的网站模板的名称，即参数 Template 的属性值。命令和输出结果如下：

SPO PS > Get-SPOWebTemplate

```
Name                        Title                          LocaleId    CompatibilityLevel
----                        -----                          --------    ------------------
STS#3                       团队网站                        2052        15
STS#0                       团队网站(经典体验)               2052        15
BLOG#0                      博客                            2052        15
BDR#0                       文档中心                        2052        15
DEV#0                       开发人员网站                    2052        15
OFFILE#1                    记录中心                        2052        15
EHS#1          工作组网站 -SharePoint Online 配置           2052        15
BICenterSite#0              商业智能中心                    2052        15
SRCHCEN#0                   企业搜索中心                    2052        15
BLANKINTERNETCONTAINER#0    发布门户                        2052        15
ENTERWIKI#0                 企业 Wiki                       2052        15
PROJECTSITE#0               项目网站                        2052        15
PRODUCTCATALOG#0            产品目录                        2052        15
COMMUNITY#0                 社区网站                        2052        15
COMMUNITYPORTAL#0           社区门户                        2052        15
SITEPAGEPUBLISHING#0        通信网站                        2052        15
SRCHCENTERLITE#0            基本搜索中心                    2052        15
vispr#0                     Visio 流程存储库                2052        15
```

2. 网站的删除

（1）在 SharePoint 管理中心的活动站点列表中单击要删除的站点名称左侧的复选框选中指定的网站，然后单击菜单栏上的"删除"按钮，删除该网站。类似地，我们可以使用 Remove-SPOSite 命令删除 SharePoint Online 网站，必选的参数是即将删除网站的 URL。下述 PowerShell 命令将删除 Blogs 网站。命令和输出结果如下：

```
SPO PS > Remove-SPOSite -Identity "https://office365ps.sharepoint.cn/sites/Blogs"

Confirm
Are you sure you want to perform this action?
Performing the operation "Remove-SPOSite" on target
"https://office365ps.sharepoint.cn/sites/Blogs".
 [Y] Yes  [A] Yes to All  [N] No  [L] No to All  [S] Suspend  [?] Help (default
is "Y"): Y
```

我们也可以为上述命令添加 Confirm 参数，使网站的删除操作不需要管理员进行确认。命令如下：

```
SPO PS > Remove-SPOSite -Identity "https://office365ps.sharepoint.cn/sites/Blogs" `
>> -Confirm:$false
```

（2）被删除的 SharePoint Online 网站会被放入已删除的网站列表。在 SharePoint 管理中心左侧的导航栏中单击"网站"按钮，并在弹出的扩展菜单中单击"已删除的网站"按钮，即可在右侧窗格中的已删除的网站列表中查看已经被删除的网站，如图 5-8 所示。

图 5-8

使用 PowerShell 的不带参数的 Get-SPODeletedSite 命令，同样可以列出回收站中所有被删除的网站。命令和输出结果如下：

SPO PS > Get-SPODeletedSite

```
Url                                                      Storage Quota Resource
Quota Deletion Time        Days Remaining
---                                                      ------------- 
--------------    --------------    --------------
  https://office365ps.sharepoint.cn/sites/Blogs               26214400
0 10/15/2018 12:59:47 PM         92
  https://office365ps.sharepoint.cn/sites/visprs              26214400
0 10/11/2018 11:18:47 PM         73
  https://office365ps.sharepoint.cn/portals/hub               26214400
0 10/11/2018 11:15:29 PM         73
  https://office365ps.sharepoint.cn/sites/sitepagepublishing  26214400
0 10/11/2018 11:19:02 PM         73
```

如果配合 Identity 参数，并将已删除网站的 URL 作为该参数的属性值，该命令可以查询被删除网站的具体信息。命令和输出结果如下：

SPO PS > Get-SPODeletedSite -Identity "https://office365ps.sharepoint.cn/sites/Blogs" | >> Format-List

```
SiteId          : 472371e6-d349-4994-8cf5-8de4d0cdb74a
Url             : https://office365ps.sharepoint.cn/sites/Blogs
Status          : Recycled
DeletionTime    : 10/15/2018 12:59:47 PM
DaysRemaining   : 92
StorageQuota    : 26214400
ResourceQuota   : 0
```

（3）在 SharePoint 管理中心的已删除的网站列表中，SharePoint Online 的管理员可以单击要还原的网站的站点名称以选择该网站，然后在菜单栏中单击"还原"按钮，还原该网站，或者在选择该网站后，单击菜单栏中的"永久删除"按钮，并在弹出的"永久删除站点"对话框中单击"删除"按钮来彻底删除该网站，如图 5-9 所示。

图 5-9

被删除的网站将在已删除的网站列表中被保留 93 天的时间，如果此时我们执行 PowerShell 的 Remove-SPODeletedSite 命令，便会将其彻底删除。网站被彻底删除后，即使我们运行还原网站的命令，也不可能再恢复该网站了。彻底删除网站的命令和输出结果如下：

SPO PS > Remove-SPODeletedSite -Identity "https://office365ps.sharepoint.cn/sites/Blogs"

```
Confirm
Are you sure you want to perform this action?
Performing the operation "Permanently removing site." on target
"https://office365ps.sharepoint.cn/sites/Blogs".
```

```
       [Y] Yes  [A] Yes to All  [N] No  [L] No to All  [S] Suspend  [?] Help (default
is "Y"): Y
```

在已删除的网站列表中的网站是不能通过浏览器进行访问的，并且在 93 天的系统保留期结束后，会被自动删除。在网站被彻底删除前，使用 Restore-SPODeletedSite 命令可以将已经删除的网站还原到未删除之前的状态。该命令必选的参数只有一个，即要被还原网站的 URL。在还原网站的过程中，恢复命令会将 SPODeletedSite 对象转换回 SPOSite 对象。例如，我们要恢复之前被删除的 Blogs 网站，只需执行下面的 PowerShell 命令：

SPO PS > Restore-SPODeletedSite -Identity "https://office365ps.sharepoint.cn/sites/Blogs"

5.2.3 如何测试与修复网站

1. 网站的测试

如果我们修改了网站的设置或者为网站部署了新版本后，网站出现访问异常甚至崩溃的情况，可以单击该网站的顶级网站页面右上角齿轮形状的"设置"按钮，在弹出的扩展菜单中单击"网站设置"按钮，如图 5-10 所示。

图 5-10

在随后弹出的"网站设置"页面中单击"网站集管理"项下的"网站集状况检查"链接，检查网站集的状况，如图 5-11 所示。

图 5-11

使用 PowerShell 的 Test-SPOSite 命令也可以实现类似的功能，必选的参数就是网站的URL。命令和输出结果如下：

SPO PS > Test-SPOSite -Identity "https://office365ps.sharepoint.cn"

```
SiteUrl              : https://office365ps.sharepoint.cn
Results              : {
                       SPSiteHealthResult Status=Passed RuleName="发生冲突的内容类
型" RuleId=befe203b-a8c0-48c2-b5f0-27c10f9e1622,
                       SPSiteHealthResult Status=Passed RuleName="自定义文件"
RuleId=cd839b0d-9707-4950-8fac-f306cb920f6c,
                       SPSiteHealthResult Status=Passed RuleName="缺少库" RuleId=
ee967197-ccbe-4c00-88e4-e6fab81145e1,
                       SPSiteHealthResult Status=Passed RuleName="缺少父级内容类型"
RuleId=a9a6769f-7289-4b9f-ae7f-5db4b997d284...}
PassedCount          : 7
FailedWarningCount   : 0
FailedErrorCount     : 0
```

要想具体查看哪些测试没有通过，可以使用下面的命令：

SPO PS> (Test-SPOSite -Identity "https://office365ps.sharepoint.cn").Results |
>> Where-Object {$_.Status -ne "Passed"}

注意

对于使用"团队网站"、"团队网站（无 Office 365 组）"或"通信网站"模板创建的网站，单击该网站的顶级网站页面右上角齿轮形状的"设置"按钮，在弹出的扩展菜单中是没有"网站设置"按钮的。如果需要访问"网站设置"页面，可以先在弹出的扩展菜单中单击"网站信息"按钮，然后在"网站信息"对话框中单击"查看所有网站设置"按钮，即可打开"网站设置"页面。

2. 网站的修复

如果网站存在问题，可以尝试使用 Repair-SPOSite 命令来进行网站的修复。这条命令可以检查健康分析器的规则，并在遇到错误的时候尝试自动对错误的网站进行修复。命令和输出结果如下：

```
SPO PS > Repair-SPOSite -Identity "https://office365ps.sharepoint.cn"
```

```
    SiteUrl                 : https://office365ps.sharepoint.cn
    Results                 : {
                              SPSiteHealthResult   Status=Passed   RuleName="Conflicting
Content Types" RuleId=befe203b-a8c0-48c2-b5f0-27c10f9e1622,
                              SPSiteHealthResult   Status=Passed   RuleName="Customized
Files" RuleId=cd839b0d-9707-4950-8fac-f306cb920f6c,
                              SPSiteHealthResult   Status=Passed    RuleName="Missing
Galleries" RuleId=ee967197-ccbe-4c00-88e4-e6fab81145e1,
                              SPSiteHealthResult Status=Passed RuleName="Missing Parent
Content Types" RuleId=a9a6769f-7289-4b9f-ae7f-5db4b997d284...}
    PassedCount             : 7
    FailedWarningCount      : 0
    FailedErrorCount        : 0
```

5.3 如何使用 PowerShell 管理 SharePoint Online 网站中的组

网站中的组，拥有对网站或对网站中内容的权限，包括完全控制、设计、编辑、读取及批准等。网站中的组按照类型又可以划分为 SharePoint 组、域组等。

5.3.1 如何查询网站中的组

要想查看网站中包含的组及其相应的权限，需要首先访问该网站的顶级网站的 URL。单击页面右上角的齿轮形状的"设置"按钮，接着在弹出的扩展菜单中单击"网站设置"按钮；然后，在"网站设置"页面单击"用户和权限"项中的"网站权限"链接。使用"团队网站"、"团队网站（无 Office 365 组）"或"通信网站"作为模板创建的网站，在"设置"扩展菜单中单击"网站权限"按钮，并在随后弹出的"权限"对话框中单击"高级权限设置"按钮，会产生与单击"网站权限"链接同样的效果。如图 5-12 所示。

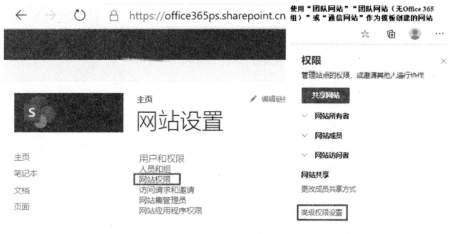

图 5-12

在接下来打开的页面（网站权限页面）中即可查看所有默认组的信息。有些网站需要单击页面提示部分的"显示用户"链接才能显示权限级别为受限访问的组的信息，如图 5-13 所示。

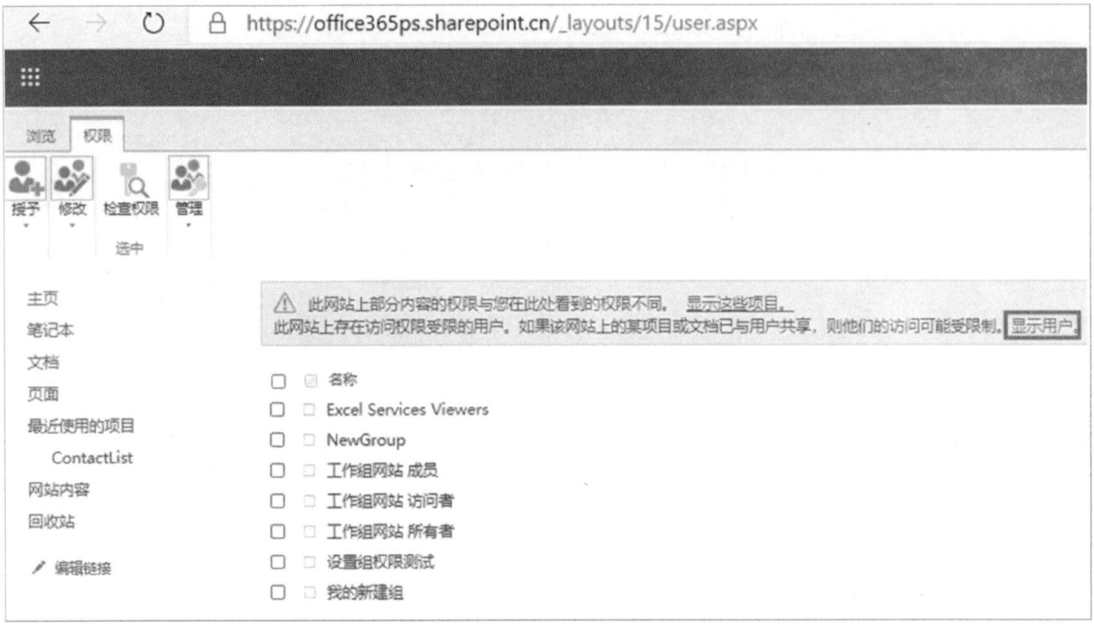

图 5-13

如果我们要查看不同网站中组的信息，需要先访问这些网站的顶级网站的 URL，并重复上述步骤。而通过使用 PowerShell 的 Get-SPOSiteGroup 命令，我们可以在同一个 PowerShell 的命令行窗口中显示这些不同网站的组列表、组中包含的用户及权限级别等信息，为管理工作提供便利。命令和输出结果如下：

```
SPO PS > Get-SPOSiteGroup -Site "https://office365ps.sharepoint.cn/search"

LoginName        : Excel Services Viewers
Title            : Excel Services Viewers
OwnerLoginName   : 所有者
OwnerTitle       : 所有者
Users            : {}
Roles            : {仅查看}

LoginName        : 层次结构管理者
Title            : 层次结构管理者
OwnerLoginName   : 所有者
OwnerTitle       : 所有者
Users            : {}
Roles            : {受限访问, 管理层次结构}

LoginName        : 成员
```

```
Title            : 成员
OwnerLoginName   : 所有者
OwnerTitle       : 所有者
Users            : {}
Roles            : {参与讨论}

LoginName        : 翻译管理器
Title            : 翻译管理器
OwnerLoginName   : 所有者
OwnerTitle       : 所有者
Users            : {}
Roles            : {翻译的限制界面}

LoginName        : 访问者
Title            : 访问者
OwnerLoginName   : 所有者
OwnerTitle       : 所有者
Users            : {}
Roles            : {读取}

LoginName        : 快速部署用户
Title            : 快速部署用户
OwnerLoginName   : 所有者
OwnerTitle       : 所有者
Users            : {}
Roles            :

LoginName        : 设计者
Title            : 设计者
OwnerLoginName   : 所有者
OwnerTitle       : 所有者
Users            : {}
Roles            : {设计, 受限访问}

LoginName        : 审批者
Title            : 审批者
OwnerLoginName   : 所有者
OwnerTitle       : 所有者
Users            : {}
Roles            : {批准}

LoginName        : 受限制读者
Title            : 受限制读者
OwnerLoginName   : 所有者
OwnerTitle       : 所有者
Users            : {}
Roles            : {受限读取}
```

```
LoginName      : 所有者
Title          : 所有者
OwnerLoginName : 所有者
OwnerTitle     : 所有者
Users          : {SHAREPOINT\system}
Roles          : {完全控制, 受限访问}

LoginName      : 样式资源读者
Title          : 样式资源读者
OwnerLoginName : 所有者
OwnerTitle     : 所有者
Users          : {true, windows}
Roles          : {受限访问}
```

不同种类的网站模板创建的 SharePoint Online 网站中的默认组有所不同，表 5-3 列出了拥有"Office 365 商业高级版"订阅，并且一开始使用的是经典 SharePoint 管理中心时，SharePoint Online 创建的三个网站中的默认组及其权限。

表 5-3

网 站	名 称	类 型	权限级别
（根）网站	Excel Services Viewers	SharePoint 组	仅查看
	工作组网站 成员	SharePoint 组	编辑
	工作组网站 访问者	SharePoint 组	读取
	工作组网站 所有者	SharePoint 组	完全控制
	设置组权限测试	SharePoint 组	读取
搜索中心	Excel Services Viewers	SharePoint 组	仅查看
	层次结构管理者	SharePoint 组	管理层次结构
	成员	SharePoint 组	参与讨论
	除外部用户外的任何人	域组	读取
	翻译管理器	SharePoint 组	翻译的限制界面
	访问者	SharePoint 组	读取
	设计者	SharePoint 组	设计
	审批者	SharePoint 组	批准
	受限制读者	SharePoint 组	受限读取
	所有者	SharePoint 组	完全控制
我的网站	成员	SharePoint 组	参与讨论
	访问者	SharePoint 组	读取
	所有者	SharePoint 组	完全控制

如果仅仅要获取某个组，例如搜索中心网站的"层次结构管理者"这个组的权限信息，我们可以使用下面的 PowerShell 命令。命令和输出结果如下：

```
SPO PS > (Get-SPOSiteGroup -Site "https://office365ps.sharepoint.cn/search" |
>> Where-Object {$_.Title -eq "层次结构管理者"}).Roles
```

受限访问
管理层次结构

上述命令首先获取搜索中心网站中所有的组,然后使用过滤条件定位"层次结构管理者"这个组,最后通过查询这个组的 Roles 属性得到"层次结构管理者"这个组的权限信息。

5.3.2 如何管理网站中的组

1. 组的创建

使用图形界面为 SharePoint Online 的网站创建新组时,需要访问该网站的顶级网站,单击页面右上角齿轮形状的"设置"按钮,在弹出的扩展菜单中单击"网站设置"按钮;在"网站设置"页面中单击"网站权限"链接,如图 5-12 所示;然后,在下一个页面的"权限"标签的菜单栏中单击"授予"按钮,并在弹出的扩展菜单中单击"创建组"按钮,如图 5-14 所示。

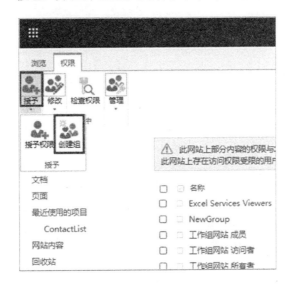

图 5-14

使用 New-SPOSiteGroup 命令则不必逐个登录每个网站的顶级网站,而可以通过 PowerShell 的控制台窗口在这些网站中直接创建组。该命令的必选参数包括网站的 URL、新组的名称及要分配给新组的权限。例如,要在根网站上创建一个名为 NewGroup 的新组,并使其具有"设计"的权限,可以使用下述命令。命令和输出结果如下:

```
SPO PS > New-SPOSiteGroup -Site "https://office365ps.sharepoint.cn/" `
>> -Group "NewGroup" -PermissionLevels "Design"
```

```
New-SPOSiteGroup : The permission level 'Design' is not a supported permission.
At line:1 char:1
```

```
    + New-SPOSiteGroup -Site https://office365ps.sharepoint.cn/ -Group "New ...
    + ~~~~~~~~~~~~~~~~~~~~~~~~~~~~~~~~~~~~~~~~~~~~~~~~~~~~~~~~~~~~~~~
    + CategoryInfo          : NotSpecified: (:) [New-SPOSiteGroup],
ArgumentException
    + FullyQualifiedErrorId :
System.ArgumentException,Microsoft.Online.SharePoint.PowerShell.NewSPOSiteGro
up
```

此处的报错是由于我们使用了与租户不一致的语言版本导致的,使用中文的权限名称即可解决上述问题。命令和输出结果如下：

SPO PS > New-SPOSiteGroup -Site "https://office365ps.sharepoint.cn/" `
>> -Group "NewGroup" -PermissionLevels "设计"

```
LoginName       : NewGroup
Title           : NewGroup
OwnerLoginName  : spoadmin@office365ps.partner.onmschina.cn
OwnerTitle      : spo Admin
Users           : {}
Roles           : {设计}
```

2. 组属性的修改

（1）如果我们要修改网站中与组权限级别有关的设置，使用图形界面时，需要在该网站顶级网站的"网站设置"页面中单击"网站权限"链接，如图 5-12 所示；然后在下一个打开的页面中，选择组名称左侧的复选框来选中特定的组，单击页面上方"权限"标签菜单栏中的"修改"按钮，并在弹出的扩展菜单中单击"编辑用户权限"按钮，如图 5-15 所示；最后，在"编辑权限"页面中修改该组的权限。

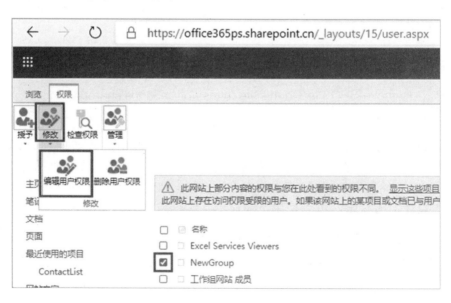

图 5-15

使用 Set-SPOSiteGroup 命令可以使我们更加高效地完成修改工作，因为修改不同网站的组权限的操作可以在同一个 PowerShell 的控制台窗口中完成。例如，下述命令把网站中的组 NewGroup 的权限级别从"设计"变更为"读取"。命令和输出结果如下：

SPO PS > Set-SPOSiteGroup -Site "https://office365ps.sharepoint.cn/" `
\>\> -Identity "NewGroup" `
\>\> -PermissionLevelsToRemove "设计" -PermissionLevelsToAdd "读取"

```
LoginName          : NewGroup
Title              : NewGroup
OwnerLoginName     : spoadmin@office365ps.partner.onmschina.cn
OwnerTitle         : spo Admin
Users              : {}
Roles              : {读取}
```

（2）另外，命令 Set-SPOSiteGroup 还可以改变组的所有者，参数 Owner 的属性值可以是用户或者安全组的标识符。命令和输出结果如下：

SPO PS > Set-SPOSiteGroup -Site "https://office365ps.sharepoint.cn/" `
\>\> -Identity "NewGroup" -Owner "admin@office365ps.partner.onmschina.cn"

```
LoginName          : NewGroup
Title              : NewGroup
OwnerLoginName     : admin@office365ps.partner.onmschina.cn
OwnerTitle         : FrankCai
Users              : {}
Roles              : {读取}
```

3. 组的删除

使用 Remove-SPOSiteGroup 命令可以删除网站中的组，命令的必选参数包括网站的 URL 及即将被删除的组的名称。命令如下：

SPO PS > Remove-SPOSiteGroup -Site "https://office365ps.sharepoint.cn/" `
\>\> -Identity "NewGroup"

如果我们提供的组名称是正确的，这条命令将不会在删除前提示用户进行确认，而是会直接删除这个组，所以要在确认组名称后再使用这条命令。

5.4 如何使用 PowerShell 管理 SharePoint Online 网站中的用户

SharePoint Online 网站中的用户可以被分为内部用户和外部用户。外部用户和外部共享是紧密相关的两个概念，只有当共享策略被设置为"内容可以与新来宾和现有来宾共享"及以上的级别时，我们才可以邀请外部用户。受邀外部用户访问网站，并且完成登录验证后，将会被添加到租户的目录中。我们可以将租户目录中的内部用户或目录中的外部用户时，添加到网站中不同的组，以授予用户在该网站中相应组的权限。另外，在网站中管理内部用户和外部用户，查看、添加及删除等操作既有相似之处，也存在不同点。

5.4.1 如何管理外部共享

1. 用户分类与外部共享

第一类是内部用户，来自 Office 365 本租户的内部，他们通常会被分配 Office 365 的许可证。

第二类是外部用户。这些用户本来不属于该 Office 365 租户，在该租户内可以由其个人或外部组织的电子邮件地址来标识，例如 xxx@oe.21vianet.com、xxx@hotmail.com 等。网站所有者或管理员可以给他们授予一个或多个网站、文件或文件夹的访问权限。虽然他们也许并不是该公司的雇员，但是想要获得对公司内部的 SharePoint Online 网站的某种级别的访问权限，进而访问存储在网站中的文档或项目文件。

外部用户还可以进一步分为：经过身份验证的具有 Microsoft 账户的用户、经过身份验证的没有 Microsoft 账户的用户和匿名的用户。其中，前两类用户的区别在于，是否拥有 Microsoft 账户或非本租户的 Office 365 账户（工作或学校账户）。如果外部用户没有 Microsoft 账户，则只能与他们共享文件或文件夹，而不能与他们共享网站。虽然我们无法验证匿名用户的身份，但是他们访问或编辑共享内容的 IP 会被记录在审核日志中。

2. 外部共享的管理

供应商和客户都可以看作企业的外部用户，使用外部共享可以提高与这些外部用户协同工作的效率，因此可以使用 SharePoint Online 的外部共享功能与外部人员共享网站、文件、或文件夹等内容。我们应该提前确定外部共享的规划，并且把其作为 SharePoint Online 整体权限规划的一部分来安排。

在默认情况下，SharePoint Online 的租户及其中各个网站的外部共享是关闭的。

（1）在 SharePoint 管理中心中单击左侧导航栏中的"策略"按钮，然后在弹出的扩展菜单中单击"共享"按钮，接着在右侧的共享属性窗格内设置本租户的共享属性，如图 5-16 所示。

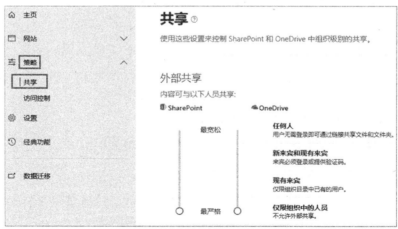

图 5-16

我们也可以使用 Set-SPOTenant 这个 PowerShell 命令来修改租户的外部共享属性，并将其修改为"允许仅与已存在于你组织目录中的外部用户共享"。命令如下：

SPO PS > Set-SPOTenant -SharingCapability ExistingExternalUserSharingOnly

修改完成后，我们再查询一下修改后的情况。命令和输出结果如下：

SPO PS > (Get-SPOTenant).SharingCapability

```
ExistingExternalUserSharingOnly
```

（2）要想改变单个网站的外部共享设置，首先需要在 SharePoint 管理中心左侧的导航栏中单击"网站"按钮，接着在弹出的扩展菜单中单击活动站点按钮，并在右侧窗格的活动站点列表中，选择该特定网站站点名称左侧的复选框，然后单击菜单栏中的"共享"按钮，如图 5-17 所示。

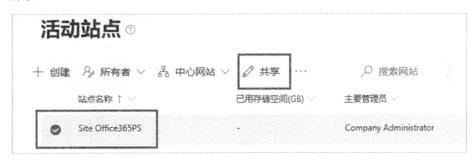

图 5-17

单击"共享"按钮后，便可以在页面右侧弹出的对话框中对该网站的外部共享属性进行设置了，如图 5-18 所示。

图 5-18

我们也可以使用下述 Set-SPOSite 命令，来修改网站的外部共享属性值。命令如下：

SPO PS > Set-SPOSite -Identity "https://office365ps.sharepoint.cn/search" `
>> -SharingCapability ExistingExternalUserSharingOnly

在 SharePoint 管理中心的活动站点列表中，可以选择多个网站，然后单击菜单栏中的"批量编辑"按钮，并在弹出的扩展菜单中单击"共享"按钮，为多个网站设置相同的外部共享属性。要想同时为不同的网站设置不相同的外部共享属性值，我们可以考虑使用 Set-SPOSite 命令，并配合 PowerShell 的循环，来便捷地完成所需进行的修改。

（3）外部共享属性和外部共享属性值是相互关联的两个不同概念，属性值定义了属性的特征，外部共享属性包括 4 个属性值，具体的含义请参考表 5-4。

表 5-4

外部共享属性值	属性值的含义
仅限组织内部人员（不允许外部共享）	用户将不能与外部用户共享网站或内容，即使这些用户已在目录当中。
仅限现有来宾（仅限组织目录中已有的来宾）	外部用户如果之前接受过邀请，或者他们已经被手动导入目录（具有#EXT#扩展名），则允许与这些已在目录中的外部用户共享网站或内容。
新的和现有来宾（来宾必须登录或提供验证码）	用户可与外部用户共享网站或内容，外部用户在查看内容之前，需要使用 Microsoft 账户登录才能访问网站或内容。
任何人（用户无须登录即可通过链接共享文件和文件夹）	除了经过身份验证的外部用户可以访问网站和内容外，受邀的收件人还允许通过匿名来宾链接，不需要登录就可以访问被共享的内容。

（4）租户外部共享属性和网站外部共享属性的设置有以下几点需要注意。

第一点，外部共享的属性值可以按照限制由少到多的顺序进行排列，网站在外部共享上的设置会被对租户范围限制更为严格的外部共享设置所代替。例如，网站之前在外部共享上的设置是 ExistingExternalUserSharingOnly，而租户此时将外部共享设置为 Disabled，则网站在外部共享上的设置也会变为 Disabled。命令和输出结果如下：

SPO PS > (Get-SPOSite -Identity https://office365ps.sharepoint.cn/search).SharingCapability

```
ExistingExternalUserSharingOnly
```

SPO PS > Set-SPOTenant -SharingCapability Disabled
SPO PS > (Get-SPOSite -Identity https://office365ps.sharepoint.cn/search).SharingCapability

```
Disabled
```

第二点，如果我们将租户的 SharingCapability 的属性值从 Disabled 重新设置为 ExistingExternalUserSharingOnly，网站也会恢复到之前可以访问的状态，即网站在外部共享上的属性值恢复为之前的 ExistingExternalUserSharingOnly。

第三点，网站在外部共享限制上设置的属性值是不能比租户级别设置的属性值更为宽松的。命令和输出结果如下：

```
SPO PS > (Get-SPOTenant).SharingCapability

ExistingExternalUserSharingOnly

SPO PS > Set-SPOSite -Identity "https://office365ps.sharepoint.cn/search" `
>> -SharingCapability ExternalUserAndGuestSharing

WARNING: The SharingCapability set for this SPOSite will not take effect until
the 'ExternalUserAndGuestSharing' SharingCapability is also enabled on the
SPOTenant.
Set-SPOSite : 不能将共享功能设置为指定的级别，因为它是比其父网站集或你的组织限制更少的设置。
At line:1 char:1
+ Set-SPOSite -Identity https://office365ps.sharepoint.cn/search -Shari ...
+ ~~~~~~~~~~~~~~~~~~~~~~~~~~~~~~~~~~~~~~~~~~~~~~~~~~~~~~~~~~~~~~~~
    + CategoryInfo          : NotSpecified: (:) [Set-SPOSite], ServerException
    + FullyQualifiedErrorId :
Microsoft.SharePoint.Client.ServerException,Microsoft.Online.SharePoint.Power
Shell.SetSite
```

5.4.2 如何管理外部用户

1. Microsoft 账户的注册

我们可以使用"邀请他人"的功能给外部用户发送邀请邮件，并通过这种方式向本租户的 Office 365 目录中添加 SharePoint Online 的外部用户。当外部用户收到邀请邮件后使用 Microsoft 账户或者其工作、学校账户的用户名和密码登录，并访问网站或内容后，系统便会将该用户添加到该 Office 365 租户的目录中，而此类账户的特征是用户名中带有"#EXT#"标识。

个人用户也可以免费注册 Microsoft 的账户。注册 Microsoft 账户的具体方法如下。

首先，在浏览器中访问微软官方网站的 URL：

https://www.microsoft.com/zh-cn/

然后，单击页面右上角的"登录"按钮，如图 5-19 所示。

图 5-19

接着，在弹出对话框中单击"创建一个！"链接，如图 5-20 所示。

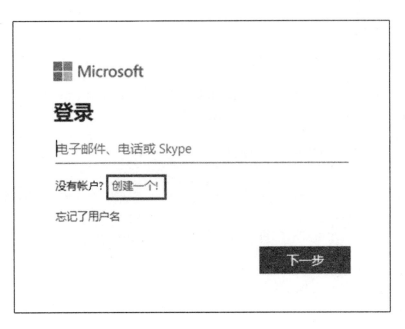

图 5-20

在弹出的下一个对话框中键入用户的非 Microsoft 账户系统的邮件地址,并单击"下一步"按钮,如图 5-21 所示。

图 5-21

在紧接着弹出的下一个对话框中键入密码(这个密码可以与用户关联 Microsoft 账户所使用的该外部邮箱的登录密码不同),并单击"下一步"按钮,如图 5-22 所示。

第 5 章　使用 PowerShell 管理 Office 365 的 SharePoint Online

图 5-22

在填写完姓、名、国籍等基本信息后，在下一个弹出的对话框中填写用户邮箱中收到的验证码，如图 5-23 所示。在完成防止机器注册的验证后，完成整个注册过程。

图 5-23

2. 外部用户的邀请

首先，单击网站（例如搜索网站）的顶级网站页面右上角齿轮形状的"设置"按钮，在弹出的扩展菜单中单击"共享…"按钮，如图 5-24 所示（直接单击"设置"按钮下方以快捷访问方式展现的"共享"按钮，也可以打开"共享网站"对话框）。

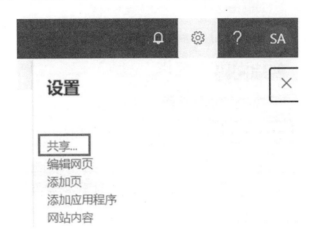

图 5-24

然后在弹出的共享网站对话框中单击左侧导航栏中的"邀请他人"按钮，并在右侧的"输入名称或电子邮件地址…"文本框中键入用户标识，即外部用户的邮件地址；最后，单击"共享"按钮，如图 5-25 所示。

图 5-25

另外，在单击"共享"按钮前，我们可以单击"显示选项"按钮（单击该按钮将显示更多选项，且"显示选项"按钮随即变为"隐藏选项"按钮），并通过"选择权限级别"下拉列表框选择将授予给外部用户的权限。

注意

对于使用"团队网站（无 Office 365 组）"或"通信网站"模板创建的网站，我们可以单击页面右上角的"设置"按钮，在打开的扩展菜单中单击"网站权限"按钮，并在页面右侧随后弹出的"权限"对话框中单击"共享网站"按钮，来打开共享网站对话框；而对于使用"团队网站"模板创建的网站，可以在"权限"对话框中单击"邀请用户"按钮，在扩展菜单中单击"仅共享网站"按钮，来打开共享网站对话框。

另外，使用以上三类模板创建的网站，虽然在共享网站对话框中无法看到"显示选项"按钮，但是当我们在添加用户、Microsoft 365 组或安全组的文本框中键入外部用户的邮件地址后，页面上即会出现为该外部用户授予权限的选项，如图 5-26 所示。

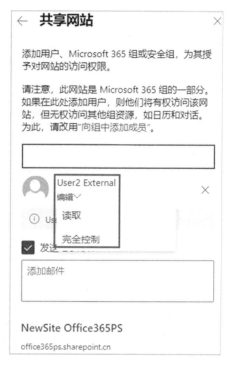

图 5-26

稍后，这个用户会在其外部邮箱中收到一封通知邮件。当用户在浏览器中访问发出邀请网站的 URL（https://office365ps.sharepoint.cn/search）时，会提示用户输入用户名，如图 5-27 所示。

图 5-27

注意

虽然我们在上述登录对话框中键入的用户名看起来与用户的外部邮件地址（仅以免费注册的 Microsoft 账户为例）相同，但是两者的含义并不相同。这里键入的其实是用户在 Microsoft 网站账户系统中注册的用户名，而不是邮件地址。

在下一个对话框中键入密码，如图 5-28 所示。这里同样也要注意，使用的密码是注册 Microsoft 网站账户系统时设定的密码，并不是用户外部邮箱的密码。

图 5-28

完成上述登录验证后，用户就可以按照网站所有者或管理员在邀请他时所授予的权限级别进行访问了。如果用户有工作或学校账户，则无须在 Microsoft 网站进行注册，登录验证过程与前述过程类似。

此外，有以下三点需要注意。

第一点，网站外部共享的属性值按照限制多少进行排序，需要将该属性值设置成"仅限现有来宾"级别以上，即"新的和现有来宾"或"任何人"，否则，当我们邀请外部用户并单击"共享"按钮后，会出现下面的警告提示，如图 5-29 所示。

图 5-29

第二点，如果用户没有 Office 365 工作或学校账户，也没有在 Microsoft 网站上注册过 Microsoft 的账户，即使邀请人完成了上述对用户的邀请步骤，用户在尝试访问该网站，并使用他的外部邮件地址登录时，也将出现没有该账户名称的报错信息，如图 5-30 所示。

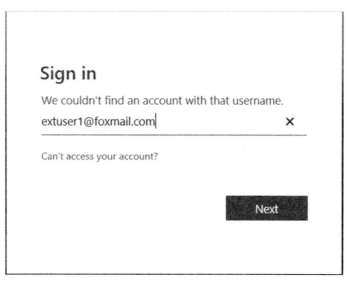

图 5-30

第三点，如果用户没有 Microsoft 账户或者 Office 365 的工作或学校账户，这些用户在每次访问时就只能使用发送给他们的具有时效性的一次性访问代码进行身份验证。并且，在用户每次需要访问共享文件或文件夹时都需要给用户发送访问代码。除非用户拥有 Microsoft 账户或者 Office 365 的工作、学校账户，否则不能与他们共享网站。

3. 外部用户的管理

如果要查看已经发送过访问邀请，而受邀用户还没有访问网站或没有成功完成登录验证的列表，当使用图形界面时，我们需要访问各个网站的顶级网站的 URL。单击页面右上角齿轮形状的"设置"按钮，在弹出的扩展菜单中单击"网站设置"按钮，并在随后的"网站设置"页面中单击"用户和权限"项下的"访问请求和邀请"链接，如图 5-31 所示。

图 5-31

在接下来打开的"访问请求"页面中，我们即可查看"外部用户邀请"项下包括的用户，如图 5-32 所示。

图 5-32

如果收到邀请的外部用户尝试访问的网站并不是邀请人最初发出邀请的那个网站，该用户进行登录验证的时候，会出现下面的报错信息，如图 5-33 所示。

图 5-33

如果要查看收到邀请且成功完成验证的用户列表，使用图形界面时需要访问各个网站的顶级网站的 URL。单击页面右上角的齿轮形状的"设置"按钮，在扩展菜单中单击"网站设置"按钮，接着在"网站设置"页面中单击"访问请求和邀请"链接，并在"访问请求"页面中单击"显示历史记录"链接，如图 5-34 所示。如果网站管理员没有从新创建的网站向外部用户发送过访问邀请，且外部用户也没有对该网站请求过访问权限，则"用户和权限"项中将不会出现"访问请求和邀请"的链接。

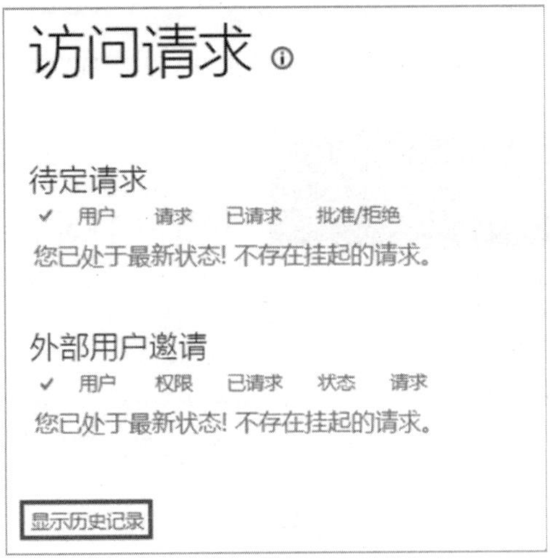

图 5-34

外部用户访问发出邀请的网站并且成功完成登录验证后，即可作为该租户用户的组成部分出现在租户的活跃用户列表中。SharePoint Online 的管理员可以登录 Office 365 管理中心，在左侧导航栏中单击"用户"按钮，并在弹出的扩展菜单中单击"活跃用户"按钮，如图 5-35 所示；然后在右侧窗格内的活跃用户列表中查看本租户内所有的用户，这其中包括受邀的外部用户。

在 Office 365 管理中心的活跃用户列表中，我们无法通过使用已有视图或创建自定义视图来达到仅查看本租户内所有 SharePoint Online 外部用户的

图 5-35

目的；同时，也无法在"搜索用户"的文本框中通过筛选外部用户的用户名中的关键字"#EXT#"，来获取所有外部用户的列表；而使用 PowerShell 的 Get-SPOExternalUser 命令来查询 Office 365 本租户中外部用户的列表，则更加方便快捷，尤其是当活跃用户列表中的用户数量比较多的时候。使用 PowerShell 命令，除了可以使用升序或降序排列用户，还可以使用参数直接找到目标对象。

使用 Get-SPOExternalUser 命令时不带任何参数，会按照时间顺序，列出最后一个收到过邀请，并且成功完成登录验证的外部用户。命令和输出结果如下：

SPO PS > Get-SPOExternalUser

```
Email         : extuser2@aliyun.com
DisplayName   : User2 External
UniqueId      : 10033230C514A1BE
AcceptedAs    : extuser2@aliyun.com
WhenCreated   : 5/18/2019 6:29:07 AM
InvitedBy     :
```

要从租户的目录中获取其他的外部用户，我们可以添加 Position 和 PageSize 参数。其中，Position 是查询的起始位置，而 PageSize 是要查询的外部用户数量的上限。命令和输出结果如下：

SPO PS > Get-SPOExternalUser -Position 0 -PageSize 15

```
Email         : extuser2@aliyun.com
DisplayName   : User2 External
UniqueId      : 10033230C514A1BE
AcceptedAs    : extuser2@aliyun.com
WhenCreated   : 5/18/2019 6:29:07 AM
InvitedBy     :

Email         : extuser1@foxmail.com
DisplayName   : User1 External
UniqueId      : 10033230C50BE4EE
```

```
AcceptedAs     : extuser1@foxmail.com
WhenCreated    : 4/15/2019 6:21:17 AM
InvitedBy      :
```

要查询本租户的目录中某个具有特定邮件地址的外部用户，我们可以添加 Filter 参数。命令和输出结果如下：

SPO PS > Get-SPOExternalUser -Filter "extuser1@foxmail.com"

```
Email          : extuser1@foxmail.com
DisplayName    : User1 External
UniqueId       : 10033230C50BE4EE
AcceptedAs     : extuser1@foxmail.com
WhenCreated    : 4/15/2019 6:21:17 AM
InvitedBy      :
```

在管理外部用户时，要注意以下几点：

（1）如果要获取在特定网站中已经被授予了访问权限的所有外部用户的列表，使用图形界面时，虽然可以在该网站的顶级网站的"网站设置"页面中单击"访问请求和邀请"链接，并在随后的"访问请求"页面中单击"显示历史记录"链接得到列表，但是该结果有可能是不准确的，因为即使是被删除的外部用户，仍然会出现在历史记录中。为了得到准确的结果，我们可以在该网站的顶级网站的"网站设置"页面中单击"网站权限"链接，然后逐个查看各个组中的成员，并在汇总后得到完整的列表。而使用 PowerShell 的 Get-SPOExternalUser 命令，则可以通过添加 SiteUrl 参数来方便地实现上述需求。命令和输出结果如下：

SPO PS > Get-SPOExternalUser -SiteUrl "https://office365ps.sharepoint.cn/search" `
>> -Position 0 -PageSize 15

```
Email          : extuser2@aliyun.com
DisplayName    : User2 External
UniqueId       : 10033230C514A1BE
AcceptedAs     : extuser2@aliyun.com
WhenCreated    : 5/18/2019 6:29:07 AM
InvitedBy      : spoadmin@office365ps.partner.onmschina.cn
```

（2）外部用户收到邀请并登录验证成功，被添加到租户的目录中后，该租户内不同网站的所有者或管理员就可以像操作内部用户一样，使用该外部用户的邮件地址作为标识在相应的网站中为该用户授予访问的权限。

（3）对于已经添加到租户目录中的外部用户，虽然 SharePoint Online 的管理员可以在 Office 365 管理中心的活跃用户列表中查看到该外部用户，但是却无法进行删除，而只能使用 PowerShell 的 Remove-SPOExternalUser 命令来进行该外部用户的删除操作。这条命令的 UniqueIDs 参数是必选的。要注意的是，该参数的值并不是用户的外部邮箱地址，而是该用户的唯一标识。所以，我们一般使用下面的方法来删除 SharePoint Online 的外部用户：先使用 Get-SPOExternalUser 命令配合 Filter 参数，获取这个外部用户的 UniqueId，然后将该用户删除。命令和输出结果如下：

```
SPO PS > Remove-SPOExternalUser `
>> -UniqueIDs (Get-SPOExternalUser -Filter "extuser2@aliyun.com").UniqueId
Confirm
Are you sure you want to perform this action?
Performing the operation "Remove external users." on target "UniqueIDs".
[Y] Yes  [A] Yes to All  [N] No  [L] No to All  [S] Suspend  [?] Help (default
is "Y"): Y
Successfully removed the following external users
10033230C514A1BE
```

（4）如果我们成功删除了租户目录中的某个外部用户，当再次尝试在网站中邀请该用户的时候，虽然共享网站的对话框在键入该用户的标识后会提示该用户在组织的外部，但是当我们单击"共享"按钮后会出现无法解析用户的警告，如图 5-36 所示。

图 5-36

5.4.3 如何管理网站中的用户

1. 网站中的管理员用户

在 SharePoint 管理中心中单击左侧导航栏中的"网站"按钮，然后在弹出的扩展菜单中单击"活动站点"按钮，并在右侧窗格内的活动站点列表中选择要修改的所有者或管理员的网站站点名称左侧的复选框，以选择该网站，然后在列表上方的菜单栏中单击"权限"按钮，如图 5-37 所示。

图 5-37

接着,在弹出的"权限"扩展菜单上,单击"管理管理员"按钮,修改网站的管理员或者主要管理员,如图 5-38 所示。对于以"团队网站"为模板创建的网站,在扩展菜单上包括两个按钮:Manage group owners 按钮用来管理组的所有者,Mange additional admins 按钮用来管理组的所有者以外的其他管理员。

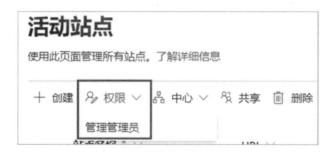

图 5-38

以"更改管理员"为例,我们可以在页面右侧弹出的"管理管理员"对话框中查看网站当前的管理员,或者通过在"添加管理员"文本框中键入用户的电子邮件地址或部分用户名,来查找并添加网站的管理员,如图 5-39 所示。

图 5-39

单一网站可以有多个具有管理员权限的用户，例如此处的 Company Administrator 和 SharePoint Service Administrator，但是最多只能有一个主要管理员，也就是网站的所有者，该网站的所有者是 Company Administrator。我们可以单击管理员名称右侧的"（×）删除"按钮来删除该网站的管理员，除非该管理员同时也是该网站的主要管理员。

（1）使用Set-SPOUser命令同样可以方便地为网站添加或删除管理员，例如，以下命令为user 02分配根网站的网站管理员角色。命令和输出结果如下：

```
SPO PS > Set-SPOUser -Site "https://office365ps.sharepoint.cn" `
>> -LoginName "user02@office365ps.partner.onmschina.cn" `
>> -IsSiteCollectionAdmin $true
```

```
Display Name  Login Name                                    Groups  User Type
------------  ----------                                    ------  ---------
user 02       user02@office365ps.partner.onmschina.cn  {}     Member
```

（2）我们可以使用 Get-SPOUser 命令查看特定网站包含哪些具有管理员权限的用户，而下述命令执行的结果中就包括被分配了网站管理员角色的用户 user 02。命令和输出结果如下：

```
SPO PS > Get-SPOUser -Site "https://office365ps.sharepoint.cn" |
>> Where-Object {$_.IsSiteAdmin -eq $true} | Select-Object LoginName, IsSiteAdmin
```

```
LoginName                                      IsSiteAdmin
---------                                      -----------
eb09cf2a-6c70-4e5b-ad82-8cc01c8b97ab           True
4e57819e-0bb6-40d9-93f6-d324d3f113c9           True
user02@office365ps.partner.onmsch...           True
```

命令输出结果中的另外两个登录名是用GUID表示的，我们可以进一步查询这两个GUID分别对应哪两个角色。我们新打开一个PowerShell的控制台窗口，使用Connect-MsolService命令，以spo Admin用户的登录凭据来连接Office 365，并运行下面的命令。命令和输出结果如下：

```
SPO PS > $cred = Get-Credential
```

```
cmdlet Get-Credential at command pipeline position 1
Supply values for the following parameters:
Credential
```

```
SPO PS > Connect-MsolService -Credential $cred -AzureEnvironment AzureChinaCloud
SPO PS > Get-MsolRole -ObjectId "eb09cf2a-6c70-4e5b-ad82-8cc01c8b97ab" |
>> Select-Object ObjectId,Name
```

```
ObjectId                              Name
--------                              ----
eb09cf2a-6c70-4e5b-ad82-8cc01c8b97ab  Company Administrator
```

```
SPO PS > Get-MsolRole -ObjectId "4e57819e-0bb6-40d9-93f6-d324d3f113c9" |
>> Select-Object ObjectId,Name
```

```
ObjectId                                 Name
--------                                 ----
4e57819e-0bb6-40d9-93f6-d324d3f113c9 SharePoint Service Administrator
```

从上述命令的输出结果中可以看到，这两个GUID分别是Company Administrator和SharePoint Service Administrator这两个系统角色。我们还可以进一步利用Get-MsolRoleMember命令，查询我们分别给哪些用户分配过这两个管理员的角色。命令和输出结果如下：

SPO PS > Get-MsolRoleMember -RoleObjectId "eb09cf2a-6c70-4e5b-ad82-8cc01c8b97ab"

```
RoleMemberType EmailAddress                              DisplayName isLicensed
-------------- ------------                              ----------- ----------
User           admin@office365ps.partner.onmschina.cn    FrankCai    True
User           user36@office365ps.partner.onmschina.cn   user 36     True
```

SPO PS > Get-MsolRoleMember -RoleObjectId "4e57819e-0bb6-40d9-93f6-d324d3f113c9"

```
RoleMemberType EmailAddress                              DisplayName isLicensed
-------------- ------------                              ----------- ----------
User           spoadmin@office365ps.partner.onmschina.cn spo Admin   True
```

（3）查看网站的主要管理员，即所有者，我们会用到之前介绍过的 Get-SPOSite 命令。命令和输出结果如下：

SPO PS > (Get-SPOSite -Identity "https://office365ps.sharepoint.cn").Owner

```
eb09cf2a-6c70-4e5b-ad82-8cc01c8b97ab
```

2. 网站中用户的查询

用户在网站中的权限依赖于其所加入的组。使用图形界面查询网站中的用户、用户所属的组及相应组的权限等信息时，我们需要访问各个网站的顶级网站，单击页面右上角的齿轮形状的"设置"按钮，在弹出的扩展菜单中单击"网站设置"按钮，然后在"网站设置"页面中单击"网站权限"链接，并在接下来打开的页面中通过单击不同组的名称，来获取对应组中所包含的用户等信息，如图 5-40 所示。

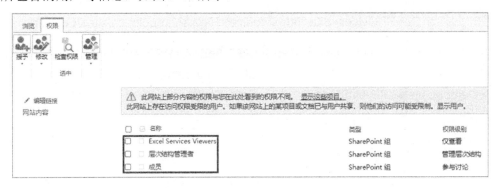

图 5-40

在网站权限页面中，我们需要逐一查看各个 SharePoint 组中的用户或安全组，并在汇总后，才能获得在该网站中拥有权限的全部用户和安全组的列表。

而通过使用 PowerShell 的 Get-SPOUser 命令，我们可以更加方便地获取该列表，因为该命令可以直接查询在该网站中拥有权限的用户（包括内部用户和外部用户）和安全组。这条命令包括必选参数 Site，参数值是网站的 URL。命令和输出结果如下：

SPO PS > Get-SPOUser -Site "https://office365ps.sharepoint.cn/search"

```
Display Name            Login Name                                      Groups  User Type
------------            ----------                                      ------  ---------
CHN\_spocachefull       chn\_spocachefull                               {}      Member
CHN\_spocrawler_3_24    chn\_spocrawler_3_24                            {}      Member
Company Administrator   eb09cf2a-6c70-4e5b-ad82-8cc01c8b97ab            {}      Member
FrankCai                admin@office365ps.partner.onmschina.cn          {成员} Member
SharePoint Service...   4e57819e-0bb6-40d9-93f6-d324d3f113c9            {}      Member
spo Admin               spoadmin@office365ps.partner.onmsch...          {}      Member
User2 External          extuser2_aliyun.com#ext#@office365...           {成员} Guest
除外部用户外的任何人    spo-grid-all-users/6578fce9-2a78-4b...          {}      Member
每个人                  true                                            {样...} Member
所有用户(windows)       windows                                         {样...} Member
系统账户                SHAREPOINT\system                               {所...} Member
```

在命令的输出结果中包含一些系统账户，例如 CHN_spocachefull、CHN_spocrawler_3_24 等，为了保证 SharePoint Online 的正常运行，这些账户都是必需的。另外，外部用户的用户类型显示为 Guest。

如果执行 Get-SPOUser 命令的用户不是特定网站的网站管理员，即使该用户是本租户的全局管理员，也会出现下述权限不足的报错信息。命令和输出结果如下：

O365 PS > $cred= Get-Credential

```
cmdlet Get-Credential at command pipeline position 1
Supply values for the following parameters:
Credential
```

O365 PS > $uri = 'https://office365ps-admin.sharepoint.cn'
O365 PS > Connect-SPOService -Url $uri -Credential $cred
O365 PS > Get-SPOUser -Site "https://office365ps.sharepoint.cn/sites/Entwiki"

```
Get-SPOUser : Access denied. You do not have permission to perform this action
or access this resource.
At line:1 char:1
+ Get-SPOUser -Site "https://office365ps.sharepoint.cn/sites/Entwiki"
+ ~~~~~~~~~~~~~~~~~~~~~~~~~~~~~~~~~~~~~~~~~~~~~~~~~~~~~~~~~~~~~~~~~~
    + CategoryInfo          : NotSpecified: (:) [Get-SPOUser],
ServerUnauthorizedAccessException
    + FullyQualifiedErrorId :
Microsoft.SharePoint.Client.ServerUnauthorizedAccessException,Microsoft.Onlin
e.SharePoint.PowerShell.GetSPOUser
```

注意

前面介绍过的 Get-SPOExternalUser 命令和这里使用的 Get-SPOUser 命令的主要区别是，前者查询到的对象只包含外部的用户，如果不指定网站，查询的范围是整个租户；而后者查询到的对象不仅包括内部、外部的用户，还包括安全组，查询的范围仅是特定的网站。

3. 网站中用户的添加

（1）将本租户的内部用户添加到网站不同的组中，即授予用户在该网站中相应组的权限。使用图形界面可以在顶级网站的"网站设置"页面中单击"网站权限"链接，并在接下来打开的网站权限页面中单击网站中组的名称。例如要将用户添加到"成员"组中，我们就单击名称为"成员"的组的链接，如图 5-40 所示。在随后打开的人员和组页面中单击菜单栏中的"新建"按钮，并在弹出的扩展菜单中单击"将用户添加到此组"按钮，如图 5-41 所示。

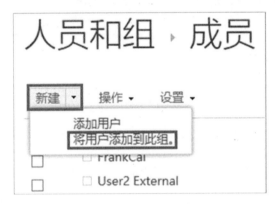

图 5-41

使用 PowerShell 的 Add-SPOUser 命令，我们不必访问每个网站的顶级网站中的网站权限页面即可将用户添加到这些网站中相应的组。这条命令需要三个参数，分别是网站的 URL、为用户指定的 SharePoint Online 组的组名称及用户的登录名，即用户的邮件地址。命令和输出结果如下：

```
SPO PS > Add-SPOUser -Site "https://office365ps.sharepoint.cn/search" `
>> -LoginName "user06@office365ps.partner.onmschina.cn" -Group "审批者"

Display Name  Login Name                                     Groups    User Type
------------  ----------                                     ------    ---------
user 06       user06@office365ps.partner.onmschina.cn        {审批者}  Member
```

（2）收到邀请的外部用户成功完成指定网站的登录验证后，会被添加到租户的目录中，这也是为该外部用户在本租户的其他网站授权的前提条件。如果该外部用户尝试访问该租户中的其他网站，将会收到权限不足的提示。此时，该用户可以单击页面中的"请求访问"按钮，来申请其他网站的访问权限，如图 5-42 所示。

第 5 章　使用 PowerShell 管理 Office 365 的 SharePoint Online

> 您需要访问此网站的权限。
>
> 我希望访问。
>
> 请求访问

图 5-42

如果访问请求发送成功，该外部用户将会在页面中看到访问请求正在等待批准的提示。而收到该访问请求的网站管理员可以使用下面的步骤查看并审批用户的访问请求。首先，网站管理员在该网站的顶级网站的"网站设置"页面中单击"访问请求和邀请"链接，并在"访问请求"页面中的"待定请求"列表中单击"批准"或"谢绝"按钮来批复用户的请求，如图 5-43 所示。而且，网站的管理员可以在批准用户的请求后，继续编辑该外部用户的权限。

访问请求

待定请求

✓	用户	请求	已请求	批准/拒绝
	User2 External	Site Office365PS	13 分钟前	批准　谢绝

图 5-43

而使用 PowerShell 的 Add-SPOUser 命令，则可以按照公司对网站的权限规划直接为租户目录中的外部用户授予租户内不同网站的访问权限，使外部用户可以毫无障碍地访问相应的资源。命令和输出结果如下：

```
SPO PS > Add-SPOUser -Site "https://office365ps.sharepoint.cn/" `
>> -LoginName "extuser2@aliyun.com" -Group "工作组网站 访问者"

Display Name    Login Name                                                          Groups
User Type
------------    ----------                                                          ------
----------
User2 External  extuser2_aliyun.com#ext#@office365ps.partner.onmschina.cn {工
作组网站 访问者} Guest
```

尚未被添加到租户目录中的外部用户，系统是无法识别的，我们也无法将其添加到特定网站中的组,并为其分配权限。例如,在外部用户 extuser3@163.com 尚未被添加到本租户的目录以前,尝试把该用户添加到网站的"成员"组,将会出现"找不到指定的用户"的报错。命令和输出结果如下:

SPO PS > Add-SPOUser -Site "https://office365ps.sharepoint.cn/search" `
>> -LoginName "extuser3@163.com" -Group "成员"

```
Add-SPOUser : 找不到指定的用户 extuser3@163.com。
At line:1 char:1
+ Add-SPOUser -Site https://office365ps.sharepoint.cn/search -LoginName ...
+ ~~~~~~~~~~~~~~~~~~~~~~~~~~~~~~~~~~~~~~~~~~~~~~~~~~~~~~~~~~~~~~~~~~~~~
    + CategoryInfo          : NotSpecified: (:) [Add-SPOUser], ServerException
    + FullyQualifiedErrorId : 
Microsoft.SharePoint.Client.ServerException,Microsoft.Online.SharePoint.Power
Shell.AddSPOUser
```

4. 网站中用户的删除

(1) 将本租户的内部用户从网站的组中删除,即移除该用户在该网站中的相应权限。首先在"网站设置"页面中单击"网站权限"链接,然后在打开的权限页面中单击网站中组的名称。例如用户 user 06 当前是"成员"组中的一员,则单击名称为"成员"的组的链接;然后,在人员和组页面的组成员列表中选择用户 user 06 左侧的复选框,单击菜单栏中的"操作"按钮,并在弹出的扩展菜单中单击"从用户组中删除用户"按钮,如图 5-44 所示。

图 5-44

使用 PowerShell 的 Remove-SPOUser 命令更加简单快捷,可以直接从指定的网站中删除具有访问权限的用户。该命令所需的参数包括网站的 URL 及被删除用户的登录名称。命令如下:

SPO PS > Remove-SPOUser -Site "https://office365ps.sharepoint.cn/search" `
>> -LoginName "user06@office365ps.partner.onmschina.cn"

第 5 章 使用 PowerShell 管理 Office 365 的 SharePoint Online

（2）将外部用户从网站的组中删除，与之前介绍过的从网站中删除本租户的用户的步骤类似，请参考图 5-44。

使用 Remove-SPOUser 命令从特定的网站中删除外部用户时，如果使用外部用户的邮件地址作为标识，将会出现"无法完成此操作"报错信息。命令和输出结果如下：

```
SPO PS > Remove-SPOUser -Site "https://office365ps.sharepoint.cn/search" `
>> -LoginName "extuser2@aliyun.com"
```

```
Remove-SPOUser ：无法完成此操作。
请重试。
At line:1 char:1
+ Remove-SPOUser -Site https://office365ps.sharepoint.cn/search -LoginN ...
+ ~~~~~~~~~~~~~~~~~~~~~~~~~~~~~~~~~~~~~~~~~~~~~~~~~~~~~~~~~~~~~~~~~~~~
    + CategoryInfo          : NotSpecified: (:) [Remove-SPOUser],
ServerException
    + FullyQualifiedErrorId                                        :
Microsoft.SharePoint.Client.ServerException,Microsoft.Online.SharePoint.Power
Shell.RemoveSPOUser
```

上述错误的产生，是由于外部用户在网站中的登录名并不是该用户的外部邮件地址，而是包含"#EXT#"标识的字符串，且用户邮件地址中的"@"符号也会被"_"符号所代替。相应地，我们可以搭配使用 Get-SPOExternalUser 命令来获取外部用户的登录名，为我们的删除操作提供进一步的方便。命令如下：

```
SPO PS > $dn = (Get-SPOExternalUser -Position 0 -PageSize 15 |
>> Where-Object {$_.Email -eq "extuser2@aliyun.com"}).DisplayName
SPO PS > Remove-SPOUser -Site "https://office365ps.sharepoint.cn/search" `
>> -LoginName (Get-SPOUser -Site "https://office365ps.sharepoint.cn/search" |
>> Where-Object {$_.DisplayName -eq $dn}).LoginName
```

注意

前面介绍过的 Remove-SPOExternalUser 命令和这里使用的 Remove-SPOUser 命令的主要区别是，前者是从租户的目录中删除用户，使用外部用户的 UniqueId 作为删除标识；而后者仅在特定网站的范围内将内部、外部用户或安全组从网站中移除，使用登录名作为删除的标识。

（3）受邀的外部用户在完成网站的登录验证，被添加到租户的目录后，网站的所有者或管理员可以利用该用户的邮件地址作为标识在不同的网站中，为该用户授予权限。如果我们要彻底删除租户目录中的某个外部用户及该外部用户在租户内各个网站的权限，就需要注意以下两点。

首先，因为在 SharePoint 管理中心中无法统一管理特定的外部用户在租户内各个网站中的权限，我们就需要访问租户内每个网站顶级网站的"权限"页面，然后在各个组的成员列表中进行逐一的排查与删除操作。而通过使用 Remove-SPOUser 命令，并配合 PowerShell 的循环结构，可以降低删除操作的烦琐程度。

其次，为了将租户目录中的某个外部用户彻底删除，我们可以使用Remove-SPOExternalUser命令。虽然既要用到Remove-SPOExternalUser命令，也要用到Remove-SPOUser命令，但是如果先使用Remove-SPOExternalUser命令从租户的目录中删除该外部用户，在各个网站给该外部用户授予的访问权限并不会被同步删除。命令和输出结果如下：

```
SPO PS > Remove-SPOExternalUser `
>> -UniqueIDs (Get-SPOExternalUser -Filter "extuser1@foxmail.com").UniqueId
```

```
Confirm
Are you sure you want to perform this action?
Performing the operation "Remove external users. " on target "UniqueIDs".
[Y] Yes  [A] Yes to All  [N] No  [L] No to All  [S] Suspend  [?] Help (default
is "Y"): Y
Successfully removed the following external users
10033230C50BE4EE
```

```
SPO PS > Get-SPOUser -Site "https://office365ps.sharepoint.cn/search" |
>> Where-Object {$_.LoginName -match "extuser1"}
```

```
Display Name    Login Name                                                Groups    User Type
------------    ----------                                                ------    ---------
User1 External  extuser1_foxmail.com#ext#@office365ps.partner.onmschina.cn {成员}    Guest
```

综上所述，彻底删除指定外部用户的推荐步骤是先使用Get-SPOUser命令遍历本租户的每个网站，如果存在对该外部用户的授权，就使用Remove-SPOUser命令将其删除；然后使用Remove-SPOExternalUser命令把该用户从租户的目录中彻底删除。命令如下：

```
SPO PS > $DelUserEmail = "extuser1@foxmail.com"
SPO PS > $DelUserId = Split($DelUserEmail,"@")[0]
SPO PS > $SiteCols = Get-SPOSite
SPO PS > foreach ($SiteCol in $SiteCols) {
>> $bingo = Get-SPOUser -Site $SiteCol.Url |
>> Where-Object {$_.LoginName -match $DelUserId}
>> if ($bingo -ne $null) {
>> Remove-SPOUser -Site $SiteCol.Url -LoginName $bingo.LoginName}
>> }
SPO PS > Remove-SPOExternalUser `
>> -UniqueIDs (Get-SPOExternalUser -Filter $DelUserEmail).UniqueId
```

5.5 如何使用 PowerShell 管理 OneDrive for Business 个人网站

OneDrive 是联机个人存储服务，而当我们购买了 Office 365 的订阅后，就可以使用 OneDrive for Business，后者包括了一些高级 OneDrive 的功能和 1TB 默认容量的存储空间。OneDrive for Business 承担了 SharePoint Online 个人网站的角色，二者有着千丝万缕的联系。OneDrive for Business 界面的展示方式和设置方式是从 SharePoint Online 网站继承而来的，OneDrive for Business 网站的 URL 中总是包含 https://office365ps-my.sharepoint.cn/部分，如图 5-45 所示。

图 5-45

另外，管理 OneDrive for Business 网站所使用的 PowerShell 命令也包含 SPO 的标识，是 SharePoint Online Management Shell 模块的子集。多个管理 SharePoint Online 的 PowerShell 命令在添加相应的参数后，也是管理 OneDrive for Business 的常用命令。

5.5.1 如何查看个人网站

虽然在 SharePoint 管理中心的活动站点列表中，我们无法查看 OneDrive for Business 个人网站，但是如果用户已经访问过他们的 OneDrive for Business 个人网站了，我们则可以通过以下方法查看用户个人网站的 URL。首先，在 SharePoint 管理中心左侧的导航栏中单击"更多功能"按钮；然后，在右侧的"更多功能"窗格内单击"用户个人资料"项下的"打开"按钮；接着，在弹出的"用户配置文件"页面中的"人员"标签下，单击"管理用户配置文件"链接，如图 5-46 所示。

图 5-46

在进行用户配置文件管理的页面中，在"查找配置文件"文本框中键入用户标识，例如用户的邮件地址，然后单击"查找"按钮。如果该用户的配置文件存在并被显示在下方的列表中，用鼠标右键单击该用户的账户名，并在弹出的快捷菜单上单击"编辑我的个人资料"按钮，如图 5-47 所示。

图 5-47

如果该用户有 OneDrive for Business 的个人网站，在该用户的"用户配置文件页面"的"个人网站"项下，就会显示该用户个人网站的部分 URL，如图 5-48 所示。

图 5-48

为了避免通过逐个查看用户的配置文件汇总得到用户个人网站列表的麻烦，我们可以使用之前介绍过的 PowerShell 的 Get-SPOSite 命令，并配合 IncludePersonalSite 参数，来查询用户的 OneDrive for Business 个人网站。命令和输出结果如下：

```
SPO PS > Get-SPOSite -IncludePersonalSite $true |
>> Where-Object {$_.Url -match "https://office365ps-my.sharepoint.cn/personal/"}

   Url
Owner
   ---
-----
   https://office365ps-my.sharepoint.cn/personal/spoadmin_office365ps_partner
_onmschina_cn       spoadmin@office365ps.par...
   https://office365ps-my.sharepoint.cn/personal/user31_office365ps_partner_o
nmschina_cn         user31@office365ps.partn...
   https://office365ps-my.sharepoint.cn/personal/admin_office365ps_partner_on
mschina_cn          admin@office365ps.partne...
   https://office365ps-my.sharepoint.cn/personal/user01_office365ps_partner_o
nmschina_cn         user01@office365ps.partn...
```

由于用户的 OneDrive for Business 个人网站的 URL 总是包含固定的字符串 https://office365ps-my.sharepoint.cn/personal/，上述命令便利用该字符串作为过滤条件，来排除非个人网站的其他网站条目。

5.5.2 如何制备个人网站

我们无法在 SharePoint 管理中心中通过创建网站的按钮来为用户制备 OneDrive for Business 的个人网站。在默认设置下，租户中的用户首次访问其个人网站的时候，系统会为这个用户触发制备个人网站的操作。由于制备过程需要一些时间，为了给用户提供更良好的使用体验，我们可以使用 Request-SPOPersonalSite 命令提前制备或批量制备个人网站。命令如下：

```
SPO PS > $uid = @("user11@office365ps.partner.onmschina.cn", `
>> "user12@office365ps. partner.onmschina.cn")
SPO PS > Request-SPOPersonalSite -UserEmails $uid -NoWait
```

使用这条命令后，并不是立即开始制备个人网站，而是会将制备请求放入队列，并在稍后由定时器执行。使用 NoWait 参数可以在同时为较多的用户制备网站的时候加快处理的进度。使用这条命令的时候还需要注意以下几点。

（1）如果用户已经有可以访问的 OneDrive for Business 个人网站，再次运行 Request-SPOPersonalSite 命令给该用户制备网站的时候不会有报错信息，而是会忽略该请求。

（2）为用户制备个人网站的前提条件是该用户需要有 SharePoint Online 的许可证，如果在用户没有许可证的情况下使用 Request-SPOPersonalSite 命令，将不会为该用户制备个人网站。

（3）虽然我们可以使用数组来存储需要制备个人网站的用户的邮件地址，并将该数组赋值给 UserEmails 参数。数组内元素的数量上限是 200 个，超过 200 个的部分可以使用管道操作符放入后续的组中，并以组为单位提交制备的请求。

5.5.3 如何管理个人网站

1. 个人网站存储配额的管理

由于在图形界面上的活动站点列表中无法显示个人网站，而且 OneDrive for Business 个人网站的存储配额与 SharePoint Online 非个人网站的存储配额是分别进行计算的，所以我们无法在 SharePoint 管理中心为个人网站设置存储配额，而只能通过使用 PowerShell 命令进行配额的设置。

（1）在租户范围管理个人网站的存储配额。

使用以前介绍过的 Get-SPOTenant 命令即可在租户范围查询个人网站的存储配额，而配额的数量单位是 MB。命令和输出结果如下：

SPO PS > Get-SPOTenant | Select-Object OneDriveStorageQuota

```
OneDriveStorageQuota
--------------------
            1048576
```

在租户范围设置个人网站的存储配额，会影响所有新建的个人网站的相应设置，但是，对于已经存在的个人网站，这条命令并不会修改这些网站现有的存储配额。命令和输出结果如下：

SPO PS > Set-SPOTenant -OneDriveStorageQuota 5242881
SPO PS > Get-SPOTenant | Select-Object OneDriveStorageQuota

```
OneDriveStorageQuota
--------------------
            5242880
```

> **注意**
>
> OneDrive for Business 个人网站的默认存储配额是 1TB，而 SharePoint Online 管理员可以最高将其设置为 5TB。如果我们使用上述命令，尝试将存储配额设置为 5242881MB，虽然并没有报错提示，但是依然不会超过 5TB 这个容量限制。如果用户的 5TB 个人网站存储容量已经接近上限，我们可以向微软公司的技术支持部门提出申请，并且在符合规定的情况下，进一步扩展容量上限。

（2）在用户层级管理个人网站的存储配额。

与在租户范围管理个人网站的存储配额的情况类似，我们可以使用以前介绍过的 Get-SPOSite 命令，在用户层级查看 OneDrive for Business 个人网站的存储配额设置。命令和输出结果如下：

SPO PS > Get-SPOSite -IncludePersonalSite $true |
\>> Where-Object {$_.Url -match "https://office365ps-my.sharepoint.cn/personal/"} |

>> Select-Object Url,StorageQuota

```
    Url
StorageQuota
    ---
------------
    https://office365ps-my.sharepoint.cn/personal/user12_office365ps_partner_
onmschina_cn         1048576
    https://office365ps-my.sharepoint.cn/personal/user15_office365ps_partner_
onmschina_cn         1048576
    https://office365ps-my.sharepoint.cn/personal/user11_office365ps_partner_
onmschina_cn         1048576
    https://office365ps-my.sharepoint.cn/personal/spoadmin_office365ps_partner
_onmschina_cn         1048576
    https://office365ps-my.sharepoint.cn/personal/user31_office365ps_partner_
onmschina_cn         1048576
    https://office365ps-my.sharepoint.cn/personal/user14_office365ps_partner_
onmschina_cn         1048576
    https://office365ps-my.sharepoint.cn/personal/admin_office365ps_partner_
onmschina_cn         1048576
    https://office365ps-my.sharepoint.cn/personal/user16_office365ps_partner_
onmschina_cn         1048576
    https://office365ps-my.sharepoint.cn/personal/user13_office365ps_partner_
onmschina_cn         1048576
    https://office365ps-my.sharepoint.cn/personal/user01_office365ps_partner_
onmschina_cn         1048576
```

在用户层级修改个人网站的存储配额时，该用户层级的设置将替代租户范围个人网站存储配额的设置。注意，在用户层级设置的个人网站存储配额值也是不能超过 5TB 限制的。命令和输出结果如下：

SPO PS > $publicurl = "https://office365ps-my.sharepoint.cn/personal/"
SPO PS > $privateurl = "user01_office365ps_partner_onmschina_cn"
SPO PS > $url = $publicurl + $privateurl
SPO PS > Set-SPOSite -Identity $url -StorageQuota 30720
SPO PS > Get-SPOSite -Identity $url | Select-Object Url,StorageQuota

```
    Url
StorageQuota
    ---
------------
    https://office365ps-my.sharepoint.cn/personal/user01_office365ps_partner_
onmschina_cn           30720
```

2. 个人网站的访问控制

（1）禁用 OneDrive for Business 的快捷访问方式。

要想通过在页面左上角单击九宫格形状的"应用启动器"按钮，并在弹出的"应用"扩展菜单中禁用 OneDrive for Business 快捷访问按钮的方法来禁止用户使用 OneDrive for Business 个人网站，我们可以在 SharePoint 管理中心左侧的导航栏中单击"设置"按钮，然后在右侧的"设置"窗格内单击页面底部的"经典设置页面"链接，并在随后弹出的"设置"页面中的"显示或隐藏应用磁贴"属性项中，选择"OneDrive 和 Office Online"项所对应的"隐藏"选项，如图 5-49 所示。最后单击该页面底部的"确定"按钮。

图 5-49

当用户注销并重新登录 Office 365 的门户网站后，单击页面左上角的"应用启动器"按钮，在弹出的"应用"扩展菜单中查看，就会发现 OneDrive 按钮已经消失，同时消失的按钮还包括 Word、Excel、PowerPoint 和 OneNote。图 5-50 所示的是 OneDrive 和 Office Online 按钮消失前的状态，而用户将无法再通过"应用"扩展菜单内的 OneDrive 按钮直接进入 OneDrive for Business 个人网站。

图 5-50

虽然不同用户的 OneDrive for Business 个人网站有不同的 URL，但是这些 URL 具有相似的特征，例如总是以 https://office365ps-my.sharepoint.cn/personal/作为开始，而 URL 后面区别不同用户的部分也有一定的构造规律。所以用户即使不通过单击"应用"扩展菜单中的 OneDrive 按钮，也有可能继续访问其个人网站。

（2）通过修改用户权限禁止用户创建新的个人网站。

为了禁止用户创建新的 OneDrive for Business 个人网站，我们可以在 SharePoint 管理中心左侧的导航栏中单击"更多功能"按钮，并在右侧的"更多功能"窗格内单击"用户个人资料"项下的"打开"按钮；然后在打开的"用户配置文件"页面中的"人员"标签下单击"管理用户权限"链接，如图 5-51 所示。

图 5-51

因为我们要禁止所有用户访问他们的个人网站，所以可以保留"除外部用户外的任何人"这个默认的组。接下来，去除权限中"创建个人网站（个人存储、新闻源和关注内容所必需的）"复选框的选择；最后，单击对话框底部的"确定"按钮，如图 5-52 所示。

图 5-52

将用户创建个人网站的权限禁用后，虽然以前没有个人网站的用户依然可以使用其用户凭据访问 https://office365ps-my.sharepoint.cn/，但是，该用户的 OneDrive for Business 个人网站却不会被正常地创建和配置，且个人网站上面的功能也无法正常地使用，如图 5-53 所示。

图 5-53

(3)禁止用户访问已经存在的个人网站。

修改用户权限后,虽然之前没有 OneDrive for Business 个人网站的用户无法创建新的个人网站,但是已经有个人网站的用户还是可以正常访问他们原有的个人网站的。为了彻底阻止这些用户对原有个人网站的访问,我们可以在 SharePoint 管理中心左侧的导航栏中单击"更多功能"按钮,在右侧的"更多功能"窗格内单击"用户个人资料"项下的"打开"按钮,并在随后弹出的"用户配置文件"页面中的"人员"标签下单击"管理用户配置文件"链接;然后,在用户配置文件管理页面中,通过用户标识查找到需要清除原有个人网站的用户;最后,在下方的列表中用鼠标右键单击该账户名,并在弹出的快捷菜单中单击"编辑我的个人资料"按钮。

由于这些用户的配置文件里保存有个人网站的设置,我们可以手工编辑他们的配置文件,并清空个人网站的属性值,以此来达到阻止这些用户访问原有个人网站的目的,请参考图 5-48。

另外,我们还可以使用 PowerShell 的 Remove-SPOSite 命令来替代对每个用户配置文件进行修改这一烦琐的操作,能够达到同样的效果。命令如下:

```
SPO PS > Get-SPOSite -IncludePersonalSite $true |
>> Where-Object {$_.Url -match "https://office365ps-my.sharepoint.cn/personal/"} |
>> ForEach-Object {Remove-SPOSite -Identity $_.Url -Confirm:$false}
```

注意

上述操作会同时将这些用户存储在个人网站上面的文件一并删除。如果用户需要保留这些文件,需要提前进行备份。

附录 A 本书所涉及的 PowerShell 知识点

1. PowerShell 的变量和属性

（1）PowerShell 的变量被用来临时保存数据或对象，变量的前缀为$，前缀后面的字符可以是数字、字母、下划线在内的任意字符，例如$1, $temp, $_。而$_比较特殊，可以看作是占位符，代表了变量的当前值，其经常出现在 Where-Object 或 foreach 循环后面的大括号中。PowerShell 的变量是大小写不敏感的，例如，$temp 和$TEMP 代表的是同一个变量。

（2）我们可以使用 "=" 为 PowerShell 的变量赋值，而下述命令分别将文本字符串 abc 和 "36" 赋给变量。赋值操作完成后，$temp2 是字符串类型的变量，而$temp3 是整数类型的变量。文本字符串需要使用单引号或双引号包围起来，否则赋值的过程中将报错。PowerShell 会将双引号当中以 "$" 开头的部分作为变量处理，并使用变量的值进行替换，而单引号中间的 "$" 符号仅当作普通的字符处理，而非变量的前缀。命令和输出结果如下：

PS > $temp1 = "abc"
PS > $temp2 = "36"
PS > $temp3 = 36
PS > $temp4 = abc

```
abc : The term 'abc' is not recognized as the name of a cmdlet, function, script
file, or operable program. Check the spelling of the name, or if a path was included,
verify that the path is correct and try again.
At line:1 char:10
+ $temp4 = abc
+          ~~~
    + CategoryInfo          : ObjectNotFound: (abc:String) [],
CommandNotFoundException
    + FullyQualifiedErrorId : CommandNotFoundException
```

（3）PowerShell 使用 "." 符号来引用对象的属性或方法，而数据或对象临时存储于变量中，同样可以使用 "." 符号来引用其属性或方法。命令和输出结果如下：

PS > $temp2.Length

2

另外，我们可以使用 Split 方法及指定的分隔字符将文本字符串进行切割。命令和输出结果如下：

```
PS > $colors = "red,orange,yellow"
PS > $color = $colors.Split(",")
PS > $color
```

```
red
orange
yellow
```

2. PowerShell 的输入和输出

（1）与其他脚本语言类似，PowerShell 的使用过程中也涉及信息的输入和输出，标准的输入输出操作是通过 PowerShell 的控制台完成的。从文件中获取数据是输入的另一种方式，我们可以使用 Get-Content 命令从文本文件中读取数据，我们还可以使用命令读取 CSV 格式文件中的内容。命令和输出结果如下：

```
PS > Get-Content .\ps_text.txt
```

```
test1
test2
test3
```

```
PS > Import-Csv .\ps_csv.csv
```

```
ID Name
-- ----
 1 user1
 2 user2
 3 user3
```

（2）PowerShell 中的管道（管道操作符以"|"表示）可以在单一命令行中容纳多条命令，而前一条命令的输出即为下一条命令的输入。虽然使用管道可以简化操作，但是有可能会由于前一条命令的输出结果，与下一条命令的参数所能接受的对象无法匹配，导致后一条命令执行失败。

使用管道配合 Where-Object 可以对输出的结果进行过滤和筛选，使结果仅显示符合筛选条件的内容。命令和输出结果如下：

```
PS > Get-Service | Where-Object {$_.Status -eq "Running"}
```

```
Status   Name              DisplayName
------   ----              -----------
Running  AdobeARMservice   Adobe Acrobat Update Service
Running  Appinfo           Application Information
Running  AudioEndpointBu...Windows Audio Endpoint Builder
Running  Audiosrv          Windows Audio
Running  BFE               Base Filtering Engine
Running  BITS              Background Intelligent Transfer Ser...
```

......

使用管道配合 Select-Object 可以对输出结果的属性字段进行排序和选择，使结果仅显示我们关注的属性字段，属性字段之间使用逗号分隔。命令和输出结果如下：

PS > Get-Service | Select-Object Name,Status

```
Name                                              Status
----                                              ------
AdobeARMservice                                   Running
AJRouter                                          Stopped
ALG                                               Stopped
AppIDSvc                                          Stopped
Appinfo                                           Running
AppMgmt                                           Stopped
```
......

使用管道配合 Sort-Object 可以将输出的结果列表依照特定属性字段的值进行正向或逆向的排序，例如，下述命令将所有的服务，按照 Name 字段属性值从大到小的顺序进行排列。命令和输出结果如下：

PS > Get-Service | Select-Object Name,Status | Sort-Object Name -Descending

```
Name                                              Status
----                                              ------
XboxNetApiSvc                                     Stopped
XboxGipSvc                                        Stopped
XblGameSave                                       Stopped
XblAuthManager                                    Stopped
xbgm                                              Stopped
WwanSvc                                           Stopped
```
......

使用管道配合 Format-List 可以查询特定内容的全部属性。命令和输出结果如下：

PS > Get-Service -ServiceName wuauserv | Format-List

```
Name                  : wuauserv
DisplayName           : Windows Update
Status                : Stopped
DependentServices     : {}
ServicesDependedOn    : {rpcss}
CanPauseAndContinue   : False
CanShutdown           : False
CanStop               : False
ServiceType           : Win32ShareProcess
```

PS > Get-History -Id 39 | Format-List

```
Id                       : 39
CommandLine              : get-winevent
ExecutionStatus          : Stopped
StartExecutionTime       : 6/27/2019 12:49:46 PM
EndExecutionTime         : 6/27/2019 12:49:51 PM
```

（3）当通过 PowerShell 的控制台输入命令的时候，如果添加多个参数，有可能会导致单一命令的长度过长。为了便于阅读，我们可以使用"`"符号（即 PowerShell 的转义字符）来将单一的命令行分隔为多行。

注意

"`"符号只能被添加到命令中的空白处以分隔行，而不能出现在单词的中间，也不能出现在被引号包裹的字符串的中间。

命令行被"`"符号分隔后，命令行的提示符将变更为">>"。命令和输出结果如下：

PS > Get-Service `
>> -Name "wuauserv"

```
Status   Name               DisplayName
------   ----               -----------
Stopped  wuauserv           Windows Update
```

例外情况是在以下符号后，即使不添加"`"符号，也可以换行：管道符号，Where-Object，ForEach-Object 和 PowerShell 循环结构等的"{"符号、条件判断的 if 语句、循环 for 语句等的"("符号。

3. PowerShell 的数组

（1）我们可以使用下面的方法创建一个字符串数组，而在 PowerShell 的控制台窗口中键入代表该数组的变量名称，即可查询该数组的各个元素。命令和输出结果如下：

PS > $array1 = @("WinRM","wuauserv")
PS > $array1

```
WinRM
wuauserv
```

（2）使用 Get-Content 和 Import-Csv 命令从文件中获取的数据，也可以用来创建一个数组。命令和输出结果如下：

PS > $array2 = Get-Content .\ps_text.txt
PS > $array2

```
test1
test2
test3
```

```
PS > $array3 = Import-Csv .\ps_csv.csv
PS > $array3
```

```
ID Name
-- ----
 1 user1
 2 user2
 3 user3
```

4. PowerShell 的循环

循环结构在脚本语言中是必不可少的，借助使用循环，我们可以将相同的操作逐个应用到对象，或者对集合中的对象进行遍历。虽然 PowerShell 的循环结构包括 For 循环、While 循环、Foreach 循环等，但是本书主要介绍 Foreach 循环。Foreach 循环又包括以下两种使用方式。

（1）ForEach-Object 命令通常出现在管道操作符的后面，且与代表变量当前值的$_搭配使用，对当前对象执行特定的操作。命令和输出结果如下：

```
PS > $array1 | ForEach-Object { Write-Host "$_ is a service"}
```

```
WinRM is a service
wuauserv is a service
```

（2）使用 foreach 关键字，我们可以遍历集合中的对象，括号中的表达式"$var in $array1"，$var 是当前的变量，$array1 是集合。命令和输出结果如下：

```
PS > foreach ($var in $array1) {
>> $Stat = (Get-Service | Where-Object {$_.Name -eq $var}).Status
>> Write-Host "$var is $Stat"}
```

```
WinRM is Stopped
wuauserv is Stopped
```

5. PowerShell 的条件判断

（1）虽然使用管道配合 Where-Object 可以进行判断，但是通过使用 IF-ELSEIF-ELSE 语句组合来进行条件判断，与应用 Where-Object 命令只能产生一个结果集不同，有可能会产生多个结果子集，并且我们可以进一步对每个条件的结果子集分别执行不同的命令。命令和输出结果如下：

```
PS > Get-Service | Where-Object {$_.Status -eq "Stopped"}
```

```
Status   Name         DisplayName
------   ----         -----------
Stopped  AJRouter     AllJoyn Router Service
```

```
Stopped  ALG                    Application Layer Gateway Service
Stopped  AppIDSvc               Application Identity
……
```

PS > Get-Service | ForEach-Object { if($_.Status -eq "Stopped") {
>> Write-Host $_.DisplayName "is Stopped"}
>> elseif($_.Status -eq "Running") {
>> Write-Host $_.DisplayName "is Running"}
>> }

```
Adobe Acrobat Update Service is Running
AllJoyn Router Service is Stopped
Application Layer Gateway Service is Stopped
Application Identity is Stopped
Application Information is Running
Application Management is Stopped
……
```

（2）条件的判断依赖于运算符，判断的结果为布尔型，即 True 或 False。而本书中主要是对字符串变量进行判断操作，而常用的关系运算符包括：

-eq：等于

-ne：不等于

-gt：大于

-lt：小于

命令和输出结果如下：

PS > Get-EventLog -LogName Application | Where-Object{$_.EntryType -ne "Information"}

```
Index Time         EntryType   Source              InstanceID Message
----- ----         ---------   ------              ---------- -------
17538 Jun 28 09:39  Error      Microsoft Securit...        100 
HRESULT:0x80070643...
17529 Jun 28 09:38  Warning    Outlook              2147483707 Outlook
disabled the following add-in(s):...
17498 Jun 28 06:21  Error      Microsoft Securit...        100 
HRESULT:0x80070643...
```

我们可以使用 match 或 like 操作符匹配字符串中的特定关键字，判断的结果同样为布尔型，即 True 或 False。like 操作符可以配合通配符"*"使用，而 match 操作符将使用正则表达式规则与字符串进行匹配。命令和输出结果如下：

PS > Get-EventLog -LogName Application |
>> Where-Object {$_.Message -like "HRESULT:*"}

```
 Index Time          EntryType  Source              InstanceID Message
 ----- ----          ---------  ------              ---------- -------
 17538 Jun 28 09:39  Error      Microsoft Securit...       100
HRESULT:0x80070643...
 17498 Jun 28 06:21  Error      Microsoft Securit...       100
HRESULT:0x80070643...
 17425 Jun 27 19:33  Error      Microsoft Securit...       100
HRESULT:0x80070643...
 ......
```

PS > Get-EventLog -LogName Application |
>> Where-Object {$_.Message -match "0x80070643"}

```
 Index Time          EntryType  Source              InstanceID Message
 ----- ----          ---------  ------              ---------- -------
 17538 Jun 28 09:39  Error      Microsoft Securit...       100
HRESULT:0x80070643...
 17498 Jun 28 06:21  Error      Microsoft Securit...       100
HRESULT:0x80070643...
 17425 Jun 27 19:33  Error      Microsoft Securit...       100
HRESULT:0x80070643...
 ......
```

通过使用逻辑运算符 and（与），or（或），not（非），xor（异或），我们可以将多个判断条件组合在一起，来实现更为复杂的判断标准。命令和输出结果如下：

PS > Get-EventLog -LogName Application |
>> Where-Object {($_.EntryType -eq "Warning") -and ($_.Message -match "Outlook")}

```
 Index Time          EntryType  Source              InstanceID Message
 ----- ----          ---------  ------              ---------- -------
 17529 Jun 28 09:38  Warning    Outlook             2147483707 Outlook
disabled the following add-in(s):...
 17482 Jun 28 05:55  Warning    Outlook             2147483707 Outlook
disabled the following add-in(s):...
 17468 Jun 28 05:51  Warning    Outlook             2147483707 Outlook
disabled the following add-in(s):...
 ......
```

6. PowerShell 的帮助系统

除了查看微软官方网站上提供的在线帮助手册，在 PowerShell 控制台窗口中也可以查询命令的使用帮助。

（1）使用 Update-Help 命令可以更新本地帮助系统的文档，命令开始执行后，PowerShell 控制台窗口的上方会出现当前更新的模块信息及更新的进度，如图 A-1 所示。

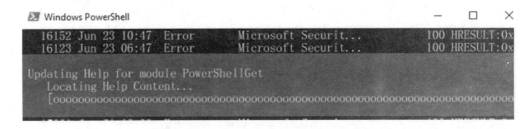

图 A-1

命令和输出结果如下：

PS > Update-Help

（2）虽然 PowerShell 的命令比较多，但是命令的名称也是有规律可循的，所以我们可以结合通配符来查询我们要使用的命令。这些命令一般会有如下的特点，以动词开始，例如 get，set，new 等；命令中包含要操作的对象信息，例如组、用户等。命令和输出结果如下：

PS > Get-Command -Name "get*user*"

```
CommandType     Name                                    Version    Source
-----------     ----                                    -------    ------
Cmdlet          Get-AzureRmADUser                       6.4.2      AzureRM.Resources
Cmdlet          Get-AzureRmApiManagementUser            6.1.5      AzureRM.ApiManagement
Cmdlet          Get-AzureRmApiManagementUserSsoUrl      6.1.5      AzureRM.ApiManagement
……
```

（3）如果我们要进一步查询命令的使用方法，我们可以使用 Get-Help 命令或 man 命令，二者的输出结果是一致的。命令和输出结果如下：

PS > Get-Help Get-LocalUser

```
NAME
    Get-LocalUser

SYNTAX
    Get-LocalUser [[-Name] <string[]>]  [<CommonParameters>]

    Get-LocalUser [[-SID] <SecurityIdentifier[]>]  [<CommonParameters>]

ALIASES
    glu

REMARKS
……
```

我们可以利用 Parameter 参数来查询该命令中具体参数的使用方法、支持类型等详细信息。命令和输出结果如下：

```
PS > Get-Help Get-LocalUser -Parameter SID
```

```
-SID <SecurityIdentifier[]>
    Required?                       false
    Position?                       0
    Accept pipeline input?          true (ByValue, ByPropertyName)
    Parameter set name              SecurityIdentifier
    Aliases                         None
    Dynamic?                        false
```

我们还可以利用 examples 参数来查询该命令的用法实例。命令和输出结果如下：

```
PS > Get-Help Get-LocalUser -Examples
```

```
NAME
    Get-LocalUser

ALIASES
    glu

REMARKS
    Get-Help cannot find the Help files for this cmdlet on this computer. It
is displaying only partial help.
        -- To download and install Help files for the module that includes this
cmdlet, use Update-Help.
        -- To view the Help topic for this cmdlet online, type: "Get-Help
Get-LocalUser -Online" or go to https://go.microsoft.com/fwlink/?LinkId=717980.
```

附录 B 本书所涉及的 PowerShell for Office 365 命令

本书章节	相关的 PowerShell for Office 365 命令
第 1 章 使用 PowerShell 管理 Office 365 的基础知识	$PSVersionTable
	Get-Module
	Install-Module
	Set-ExecutionPolicy
第 2 章 使用 PowerShell 管理 Office 365 的租户	Add-MsolGroupMember
	Add-MsolRoleMember
	Connect-MsolService
	Get-Content
	Get-Credential
	Get-MsolAccountSku
	Get-MsolDomain
	Get-MsolGroup
	Get-MsolGroupMember
	Get-MsolRole
	Get-MsolRoleMember
	Get-MsolSubscription
	Get-MsolUser
	Import-Module
	New-MsolGroup
	New-MsolLicenseOptions
	New-MsolUser
	Remove-MsolGroup
	Remove-MsolRoleMember
	Remove-MsolUser
	Restore-MsolUser

(续表)

本书章节	相关的 PowerShell for Office 365 命令
第 2 章 使用 PowerShell 管理 Office 365 的租户	Set-MsolUser
	Set-MsolUserLicense
	Set-MsolUserPassword
	Set-MsolUserPrincipalName
第 3 章 使用 PowerShell 管理 Office 365 的 Exchange Online	Add-DistributionGroupMember
	Add-MailboxPermission
	Add-RecipientPermission
	Add-UnifiedGroupLinks
	ConvertFrom-SecureString
	ConvertTo-SecureString
	Disable-Mailbox
	Enable-Mailbox
	Get-CASMailbox
	Get-Contact
	Get-Credential
	Get-DistributionGroup
	Get-DistributionGroupMember
	Get-DynamicDistributionGroup
	Get-Mailbox
	Get-MailboxPermission
	Get-MailContact
	Get-MailUser
	Get-ManagementRoleAssignment
	Get-ManagementRoleEntry
	Get-Module
	Get-MsolGroup
	Get-MsolGroupMember
	Get-MsolUser
	Get-OwaMailboxPolicy
	Get-PSSession
	Get-PublicFolder
	Get-Recipient
	Get-RecipientPermission

（续表）

本书章节	相关的 PowerShell for Office 365 命令
第 3 章 使用 PowerShell 管理 Office 365 的 Exchange Online	Get-RoleAssignmentPolicy
	Get-RoleGroup
	Get-RoleGroupMember
	Get-UnifiedGroup
	Get-UnifiedGroupLinks
	Get-User
	Import-Csv
	Import-PSSession
	New-DistributionGroup
	New-DynamicDistributionGroup
	New-Mailbox
	New-MailContact
	New-MailUser
	New-ManagementRole
	New-ManagementRoleAssignment
	New-ManagementScope
	New-Object
	New-OwaMailboxPolicy
	New-PSSession
	New-PublicFolder
	New-RoleGroup
	New-UnifiedGroup
	Out-File
	Read-Host
	Remove-DistributionGroup
	Remove-DistributionGroupMember
	Remove-DynamicDistributionGroup
	Remove-Mailbox
	Remove-MailboxPermission
	Remove-MailContact
	Remove-MailUser
	Remove-ManagementRoleAssignment
	Remove-ManagementRoleEntry

（续表）

本书章节	相关的 PowerShell for Office 365 命令
第 3 章 使用 PowerShell 管理 Office 365 的 Exchange Online	Remove-ManagementScope
	Remove-OwaMailboxPolicy
	Remove-PSSession
	Remove-RecipientPermission
	Remove-RoleGroup
	Remove-UnifiedGroup
	Remove-UnifiedGroupLinks
	Remove-UserPhoto
	Set-CASMailbox
	Set-Contact
	Set-DynamicDistributionGroup
	Set-Mailbox
	Set-MailContact
	Set-MailUser
	Set-ManagementRoleAssignment
	Set-ManagementRoleEntry
	Set-MsolUser
	Set-MsolUserLicense
	Set-MsolUserPassword
	Set-OwaMailboxPolicy
	Set-UnifiedGroup
	Set-User
	Set-UserPhoto
	Undo-SoftDeletedMailbox
	Undo-SoftDeletedUnifiedGroup
	Update-DistributionGroupMember
	Update-RoleGroupMember
	Upgrade-DistributionGroup
第 4 章 使用 PowerShell 管理 Office 365 的 Skype for Business Online	Get-Command
	Get-Credential
	Get-CsClientPolicy
	Get-CsConferencingPolicy
	Get-CsExternalAccessPolicy

（续表）

本书章节	相关的 PowerShell for Office 365 命令
第 4 章 使用 PowerShell 管理 Office 365 的 Skype for Business Online	Get-CsOnlineUser
	Get-CsPrivacyConfiguration
	Get-CsPushNotificationConfiguration
	Get-CsTenant
	Get-CsTenantFederationConfiguration
	Get-CsTenantPublicProvider
	Get-CsUserAcp
	Get-Module
	Get-PSSession
	Grant- CsExternalAccessPolicy
	Grant-CsClientPolicy
	Grant-CsConferencingPolicy
	Grant-CsMobilityPolicy
	Grant-CsPresencePolicy
	Import-Module
	Import-PSSession
	New-CsClientPolicy
	New-CsConferencingPolicy
	New-CsEdgeAllowAllKnownDomains
	New-CsEdgeAllowList
	New-CsEdgeDomainPattern
	New-CsExternalAccessPolicy
	New-CsOnlineSession
	Remove-CsClientPolicy
	Remove-CsConferencingPolicy
	Remove-CsExternalAccessPolicy
	Remove-PSSession
	Set- CsExternalAccessPolicy
	Set-CsClientPolicy
	Set-CsConferencingPolicy
	Set-CsPrivacyConfiguration
	Set-CsPushNotificationConfiguration
	Set-CsTenantFederationConfiguration

（续表）

本书章节	相关的 PowerShell for Office 365 命令
第 4 章 使用 PowerShell 管理 Office 365 的 Skype for Business Online	Set-CsTenantPublicProvider
	Set-CsUserAcp
第 5 章 使用 PowerShell 管理 Office 365 的 SharePoint Online	Add-SPOUser
	Connect-MsolService
	Connect-SPOService
	Get-Command
	Get-Credential
	Get-MsolRole
	Get-MsolRoleMember
	Get-SPODeletedSite
	Get-SPOExternalUser
	Get-SPOSite
	Get-SPOSiteGroup
	Get-SPOTenant
	Get-SPOUser
	Get-SPOWebTemplate
	Import-Module
	New-SPOSite
	New-SPOSiteGroup
	Remove-SPODeletedSite
	Remove-SPOExternalUser
	Remove-SPOSite
	Remove-SPOSiteGroup
	Remove-SPOUser
	Repair-SPOSite
	Request-SPOPersonalSite
	Restore-SPODeletedSite
	Set-SPOSite
	Set-SPOSiteGroup
	Set-SPOTenant
	Set-SPOUser
	Test-SPOSite

反侵权盗版声明

电子工业出版社依法对本作品享有专有出版权。任何未经权利人书面许可，复制、销售或通过信息网络传播本作品的行为；歪曲、篡改、剽窃本作品的行为，均违反《中华人民共和国著作权法》，其行为人应承担相应的民事责任和行政责任，构成犯罪的，将被依法追究刑事责任。

为了维护市场秩序，保护权利人的合法权益，我社将依法查处和打击侵权盗版的单位和个人。欢迎社会各界人士积极举报侵权盗版行为，本社将奖励举报有功人员，并保证举报人的信息不被泄露。

举报电话：（010）88254396；（010）88258888

传　　真：（010）88254397

E-mail：　dbqq@phei.com.cn

通信地址：北京市万寿路南口金家村 288 号华信大厦

　　　　　电子工业出版社总编办公室

邮　　编：100036